# Quality Evaluation of Plant-Derived Foods

# Quality Evaluation of Plant-Derived Foods

Editor

**Ivo Vaz de Oliveira**

MDPI • Basel • Beijing • Wuhan • Barcelona • Belgrade • Manchester • Tokyo • Cluj • Tianjin

*Editor*
Ivo Vaz de Oliveira
CITAB - Centre for the Research
and Technology of
Agro-Environmental and
Biological Sciences
University of Trás-os-Montes
and Alto Douro (UTAD)
Vila Real
Portugal


*Editorial Office*
MDPI
St. Alban-Anlage 66
4052 Basel, Switzerland


This is a reprint of articles from the Special Issue published online in the open access journal *Plants* (ISSN 2223-7747) (available at: www.mdpi.com/journal/plants/special_issues/plantderived_foods).


For citation purposes, cite each article independently as indicated on the article page online and as indicated below:

LastName, A.A.; LastName, B.B.; LastName, C.C. Article Title. *Journal Name* **Year**, *Volume Number*, Page Range.


**ISBN 978-3-0365-3548-7 (Hbk)**
**ISBN 978-3-0365-3547-0 (PDF)**

# Contents

*Article*

# Quality Control Standardization, Contaminant Detection and In Vitro Antioxidant Activity of *Prunus domestica* Linn. Fruit

Mohd Amir [1], Ameeduzzafar Zafar [2], Rizwan Ahmad [1], Wasim Ahmad [3], Mohammad Sarafroz [4], Mohammad Khalid [5], Mohammed M. Ghoneim [6], Sultan Alshehri [7], Shadma Wahab [8], Sayeed Ahmad [9] and Mohd Mujeeb [9,*]

[1] Department of Natural Products and Alternative Medicine, College of Clinical Pharmacy, Imam Abdulrahman Bin Faisal University, Dammam 31441, Saudi Arabia; matahmad@iau.edu.sa (M.A.); rareiyadh@iau.edu.sa (R.A.)

[2] Department of Pharmaceutics, College of Pharmacy, Jouf University, Sakaka 72341, Al-Jouf, Saudi Arabia; azafar@ju.edu.sa

[3] Department of Pharmacy, Mohammed Al-Mana College for Medical Sciences, Safaa, Dammam 34222, Saudi Arabia; wasimahmadansari@yahoo.com

[4] Department of Pharmaceutical Chemistry, College of Clinical Pharmacy, Imam Abdulrahman Bin Faisal University, Dammam 34212, Saudi Arabia; mskausar@iau.edu.sa

[5] Department of Pharmacognosy, College of Pharmacy, Prince Sattam Bin Abdulaziz University, P.O. Box 173, Al-Kharj 11942, Saudi Arabia; drkhalid8811@gmail.com

[6] Department of Pharmacy Practice, College of Pharmacy, AlMaarefa University, Ad Diriyah 13713, Saudi Arabia; mghoneim@mcst.edu.sa

[7] Department of Pharmaceutics, College of Pharmacy, King Saud University, Riyadh 11451, Saudi Arabia; salshehri1@ksu.edu.sa

[8] Department of Pharmacognosy, College of Pharmacy, King Khalid University, Abha 61421, Saudi Arabia; Shad.nnp@gmail.com

[9] Department of Pharmacognosy and Phytochemistry, School of Pharmaceutical Education and Research, Jamia Hamdard, New Delhi 110062, India; sahmad_jh@yahoo.co.in

* Correspondence: drmmujeeb12@gmail.com

**Citation:** Amir, M.; Zafar, A.; Ahmad, R.; Ahmad, W.; Sarafroz, M.; Khalid, M.; Ghoneim, M.M.; Alshehri, S.; Wahab, S.; Ahmad, S.; et al. Quality Control Standardization, Contaminant Detection and In Vitro Antioxidant Activity of *Prunus domestica* Linn. Fruit. *Plants* 2022, 11, 706. https://doi.org/10.3390/plants11050706

Academic Editor: Ivo Vaz de Oliveira

Received: 12 February 2022
Accepted: 28 February 2022
Published: 6 March 2022

**Publisher's Note:** MDPI stays neutral with regard to jurisdictional claims in published maps and institutional affiliations.

**Abstract:** The increase in the use of herbal medicines has led to the implementation of more stern regulations in terms of quality variation and standardization. As medicinal plants are prone to quality variation acquired due to differences in geographical origin, collection, storage, and processing, it is essential to ensure the quality, efficacy, and biological activity of medicinal plants. This study aims to standardize the widely used fruit, i.e., *Prunus domestica* Linn., using evaluation techniques (microscopic, macroscopic, and physicochemical analyses), advanced instrumental (HPLC, HPTLC, and GC–MS for phytochemical, aflatoxins, pesticides, and heavy metals), biological, and toxicological techniques (microbial load and antioxidant activities). The results revealed a 6–8 cm fruit with smooth surface, delicious odor, and acidic taste (macroscopy), thin-walled epidermis devoid of cuticle and any kind of excrescences with the existence of xylem and phloem (microscopy), LOD (15.46 ± 2.24%), moisture content (13.27 ± 1.75%), the high extractive value of 24.71 ± 4.94% in water:methanol (1:1; *v*/*v*) and with ash values in the allowed limits (physicochemical properties), and the presence of numerous phytochemical classes such as alkaloids, flavonoids, carbohydrates, glycosides, saponins, etc. (phytochemical screening). Furthermore, no heavy metals (Pb, Hg, Cd, Ar), pesticides, ad microbial limits were detected beyond the permissible limits specified, as determined with AAS, GC–MS analysis, and microbial tests. The HPTLC was developed to characterize a complete phytochemical behavior for the components present in *P. domestica* fruit extract. The parameters utilized with the method used and the results observed for the prunus herein may render this method an effective tool for quality evaluation, standardization, and quality control of *P. domestica* fruit in research, industries, and market available food products of prunus.

**Keywords:** *Prunus domestica*; standardization; GC–MS; antioxidant; HPLC; microbial limits

## 1. Introduction

The role of natural products in pharmaceutical biology is well established. Plants have been the primary source of medicine since ancient times. As per World Health Organization (WHO), >80% of the world's population of traditional medicines and used for the treatment of disease. Many drugs nowadays are either biomimetics of naturally occurring molecules or have structures derived in whole or in part from natural patterns [1]. The promotion of alternative medicine use in developed countries has been hampered by the absence of evidence of documentation and strict quality control assessments. Data of all research conducted on traditional drugs must be kept and documented as well. With this particular issue, it is essential to ensure that the herbal crude drugs and their parts used as medicine are properly standardized. There are a variety of methods and techniques that can be used in process standardization, such as pharmacognostic, phytochemical, and contaminant analyses. It is possible to identify and standardize plant material using these methods and procedures. Characterization and proper quality assurance are important steps in ensuring the quality of herbal medicine, which aids in rationalizing its safety and effectiveness [2]. Active constituents and biomarkers can be determined with reasonable precision and reliability using the HPTLC method [3]. The WHO explained conventional medicines as containing various health practices, concepts, awareness, and theories integrating herbal, mineral, and/or animal-based remedies, spiritual medication, and physical exercises performed either individually or collectively to promote health or to cure, diagnose, or prevent diseases. Some terms related to natural medicines have been provided by the WHO, according to their definitions [4].

*Prunus domestica* Linn. fruit, usually known as aloo Bukhara, belongs to the *Rosaceae* family. It is a shrubby and small deciduous tree found in Europe, Egypt, India, and Pakistan's high-altitude hilly regions of Kashmir and Swat [5]. Fresh, dried, or processed fruits from *P. domestica* have been known and consumed by humans since the ancient period. The fruit of *P. domestica* have carbohydrate, acid, and cellulose content similar to nectarines and peaches, and they all belong to the energy fruit foods [6]. Various antioxidants found in dried fruit have been shown to have anti-aging properties. Fever reducers are made from the bark of plants. Roots of *P. domestica* have astringent properties. Fatty oil from its seeds can be substituted for almond oil in many recipes [7]. *P. domestica* extract prevents fibroblast yield, enhances adrenal androgen production, and revives the secretory action of prostate and bulbourethral epithelium. There are various chemical ingredients found in *P. domestica*, i.e., flavonoids, flavonoid glycosides, abscisic acid, lignans carotenoid pigments, quinic acid, bipyrrole, and carbohydrates [8–10].

There are no previous reports available for the evaluation and standardization of *P. domestica* fruit. In the present research, the *P. domestica* fruit is subjected to various quality evaluation parameters ranging from microscopy to physicochemical, phytochemical, pharmacological, as well as toxicological analysis, in order to standardize the fruit as per the WHO guidelines. The quality evaluation may be very useful for herbal or p-pharmaceutical manufacturers and end consumers to utilize the established parameters for identification, quantification, selection, and usage of the best-quality prunus fruit.

## 2. Material and Methods

### 2.1. Reagents

Analytical grade chemicals and reagents were purchased from SD Fine Chemicals Ltd., Mumbai, India. Cadmium (Cd), lead (Pb), arsenic (As), and mercury (Hg) were obtained from Sigma Aldrich (St. Louis, MO, USA).

### 2.2. Collection and Authentication of the Drug Sample

The fruit of *P. domestica* was procured from the local market (New Delhi, India) and authenticated by a taxonomist from the Faculty of Science, Jamia Hamdard (New Delhi, India). The voucher specimen (Ref. PD/FP-369) was deposited into Herbarium and Museum, at Jamia Hamdard, New Delhi, India.

### 2.3. Macroscopic Evaluation

The macroscopical studies for the *P. domestica* were carried out visually through which the shape, color, taste, and odor were determined [11].

### 2.4. Microscopic Evaluation

For microscopic evaluation, a thin transverse section (TS) of the fruit was taken, stained with safranin (for lignification), and mounted to the microscopic slide. The sections were visualized and marked for the distinctive parts of the fruit by using a microscope.

### 2.5. Physicochemical Evaluation

The physicochemical parameters of *domestica* fruit powder—namely, loss on drying, moisture content, total water-soluble content, acid insoluble content, and extractive value, were determined. The loss on drying was analyzed at 120 °C using a hot air oven (Thermofisher, Mumbai, India), and moisture content was analyzed by Karl Fisher's titration method. The extractive value was determined by using various solvents—namely, chloroform, methanol, water: alcohol (1:1; *v/v*), and water, using the Soxhlet technique. The values were determined as per the WHO guidelines [4].

### 2.6. Phytochemical Tests (Chemical Classes Screening)

The extract of *P. domestica* fruit was subjected to different types of chemical tests to identify the presence of various natural compounds and chemical classes thereof. The phytochemical study was performed as per previously reported procedures [12].

### 2.7. Total Phenolics Content

Folin–Ciocalteu reagent method was employed for estimation of total phenolic content in *P. domestica* fruit [13]. Briefly, a dilute extract of the fruit (0.5 mL of 10 mg/mL) was mixed with Folin–Ciocalteu reagent (5 mL, 1:10 diluted with distilled water) and aqueous $Na_2CO_3$ (1 M, 4 mL). The mixture was kept for 15 min and total phenolic content was determined colorimetry at 765 nm (Shimadzu UV–Vis 1601, Tokyo, Japan) using gallic acid as standard.

### 2.8. Total Flavonoid Content

The flavonoid content of the *P. domestica* fruit was estimated by the aluminum chloride colorimetric method [13]. Briefly, the methanolic diluted extract of the fruit (0.5 mL of 10 mg/mL) was mixed with 1.5 mL of methanol, 0.1 mL of 10% aluminum chloride, 0.1 mL of 1 M potassium acetate, and 2.8 mL of distilled water. The mixture was left standing for 30 min at 25 °C, and absorbance was measurement with a UV spectrophotometer at 415 nm (Shimadzu UV–Vis 1601, Tokyo, Japan). Rutin was used as a standard for flavonoid content determination.

### 2.9. Atomic Absorption Spectrometer (AAS) Study for Evaluation of Heavy Metals

An atomic absorption spectrometer was used for determination of heavy metals such as lead (Pb), cadmium (Cd), mercury (Hg), and arsenic (Ar) in fruit powder. The instrument condition used for AAS (Model # AA-240-FS); Cd (cadmium-EDL lamp, $\lambda$ = 228.80 nm, fuel gas = acetylene at 2.5 L/min, support gas = air at 15.0 L/min), Pb (lead-EDL lamp, $\lambda$ = 283.31 nm, fuel gas = acetylene at 2.5 L/min, support gas = air at 15.0 L/min), As (arsenic-EDL lamp, $\lambda$ = 193.70 nm, fuel gas = argon at 5.5 L/min, support gas = air at 15.0 L/min), and Hg (mercury-EDL lamp, $\lambda$ = 253.65 nm, fuel gas = argon at 5.5 L/min, support gas = air at 15.0 L/min).

#### 2.9.1. Selection of Processing Parameters

The combination of fuel gas (acetylene) with subsidiary gas (air) in a mixture of 2.5: 15.0 L/min was used for the most effective separating of lead and cadmium. The mixture of 5.5:15.0 L/min of fuel gas (argon) with subsidiary gas (air) was employed as the most effective method for separating arsenic and mercury.

### 2.9.2. Optimization of the Atomic Absorption Spectra

For sharp and sensitive signals, the positive ionization mode was used for atomic absorption spectrometry detection. A standard linear calibration curve (response vs. concentration) at three different concentrations, i.e., 0.50, 1.00, and 1.50 ppm was used for optimization.

### *2.10. HPLC Determination of Aflatoxins Concentrations*
### 2.10.1. Sample Preparation

The developed methodology of association of analytical chemistry (AOAC) was used for the assessment of aflatoxins [14]. The acidic methanolic extract of *P. domestica* was neutralized with sodium chloride (NaCl) in n-hexane solution, followed by washing with dichloromethane solvent. The method was repeated 2 to 3 times to accumulate the dichloromethane layer, which was evaporated successively (2/3 mL). The resultant solvent was passed by silica gel column, followed by cleaning the column with the combination of benzene:acetic acid (9:1; *v/v*) and ether:hexane (3:1; *v/v*). The aflatoxins were eluted with 100 mL of dichloromethane:acetone (9:1; *v/v*), concentrated up to 5 mL, and desiccated with the help of inert nitrogen gas.

### 2.10.2. Derivatization of Samples (Extract) and Standards

For the derivatization of the test samples, the dried extract was dissolved in a mixture of n-hexane (200 µL) and trifluoroacetic acid (50 µL), vortexed (30 s), and left standing for 5 min. Finally, a 1.95 mL solvent system (water:acetonitrile 9:1; *v/v*) was added to the above mixture for completion of the derivatization process.

For derivatization of the standards, known concentrations (20, 40, and 80 ppb) of standard aflatoxins (B1, B2, G1, and G2) were taken and derivatized using the same aforementioned procedure for test sample derivatization.

### 2.10.3. High-Performance Liquid Chromatography (HPLC–Fluorescence) Analysis

Waters Alliance e2695 separating module (Waters, Milford, MA, USA) with C-18 column (15 cm $\times$ 4.6 mm) was applied for determination of aflatoxins. Briefly, 20 µL of the derivatized samples (standards and test samples) were injected at 1 mL/min flow rate using water:acetonitrile:methanol (70:17:17; *v/v/v*) mobile phase. A fluorescent detector was used for recording the chromatogram. For quantitative analysis, peaks of the aflatoxins were compared with standard aflatoxins peaks ($B_1$, $G_1$, $B_2$, and $G_2$).

### *2.11. GC–MS Analysis for Pesticides*

The AOAC official method (AOAC970.52/EPA525.5) was used for the determination of pesticides using GC–MS (Agilent 7890A GC system, Santa Clara, CA USA) [14]. GC–MS equipment was used (Agilent 7890A GC system, USA), and the AOAC970.52/EPA525.5 method was utilized for estimation of the pesticides. In detail, 50 mg of the sample was dissolved in methanol, followed by the addition of 50 mL of diethyl ether (added with 1 g of Na-oxalate) and petroleum ether. The mixture thus formed was vortexed for 1 min, and the organic layer was transferred into a separating funnel and added with 600 mL of water (saturated with NaCl). The aqueous layer was discarded, and this method was repeated 2 to 3 times. The resultant organic layer was mixed with sodium sulfate solution, collected, and evaporated (2–5 mL). This concentrated solution was mixed again with acetonitrile (30 mL) and petroleum ether (30 mL), passed through the column, and eluted with the help of diethyl ether. The solution obtained was concentrated (5 mL) using a rotavapor (Buchi, R-215, Flawil, Switzerland) and analyzed with the help of GC–MS.

### 2.12. HPTLC Finger Printing

For HPTLC analysis, CAMAG Linomat V (CAMAG) applicator with precoated silica gel $F_{254}$ TLC-plates (Merck, Darmstadt, Germany) was used. The extracts of *P. domestica* fruit in different solvents (chloroform, methanol, water:alcohol (1:1; *v/v*) and water) were concentrated using a rotary evaporator and dried under reduced pressure using $N_2$ gas inert condition. Finally, 5 mg/mL stock solution of each extract was prepared and spotted on TLC plates. For analysis, properly diluted stock solution was spotted via TLC plates, followed by the development of TLC plates in toluene:ethyl acetate:formic acid (5:4:0.5; *v/v/v*) solvent system for chloroform, methanol, aqueous:alcohol (1:1; *v/v*) extracts, whereas butanol:acetic acid:water (8:2:2; *v/v/v*) solvent system was used for aqueous extract TLC plate development. TLC plates were scanned by CAMAG scanner-3 at 366 nm, and photographs of the chromatograms were saved [15].

### 2.13. Microbial Contamination

The standard method as per the WHO guidelines was used for the determination of microbial load in the samples. The analysis included total fungal count, bacterial count (*Staphylococcus aureus*, *Salmonella ebony*, *Escherichia coli*, and *Pseudomonas aeruginosa*), as well other pathogen pathogens [16].

### 2.14. HPLC/DAD–DPPH Method for In Vitro Antioxidant Activity

The HPLC (Shimadzu, Kyoto, Japan) instrument, consisting of a pump (LC-10 Ai, Kyoto, Japan), a system controller (SCL-10AVP), and a diode array detector (DAD-M10 AVP), was applied for analysis of in vitro antioxidant activity of all types of extract. RP-18 column (LiChrospher®, 250 mm × 4 mm, 5 µM) (Merck, Darmstadt, Germany) was used for chromatographic separation. Methanol/water (80:20, *v/v*) used as mobile phase at 1 mL/min flow rate in isocratic elution. For data acquisition and processing, LC10 software (Version 1.6) was used.

A procedure explained by Yen and Chen was used to estimate the antioxidant potential of the samples by measuring the radical-scavenging effect of 2, 2-diphenyl-1-picrylhydrazyl (DPPH) [17]. The DPPH (100 µM) was prepared in distilled water. Briefly, DPPH solution (1 mL) was added to fruit extract sample (1 mL) and standard (ascorbic acid) of different degrees of dilution (5–100 µg/mL) in each test tube with 3 mL of distilled water, vortexed vigorously, and kept at room temperature in dark light for 10 min. The samples were filtered via a nylon membrane filter (0.2 µM), and 20 µL aliquot samples were injected for HPLC analysis. For results interpretation, the decrease in the peak area of the samples was detected at 517 nm for 10 min of run time. The differences in the decline in peak area between the control and the samples were used for calculating the % inhibition. The following formula was applied to assess the % inhibition:

$$\% \text{ inhibition} = \frac{(PA_C - PA_S)}{PA_C} \times 100$$

$PA_C$ = peak area of control, $PA_S$ = peak area of samples.

## 3. Results and Discussion

### 3.1. Macroscopical Observations

Establishing the authenticity, identity, and purity of a medicinal plant can start with an organoleptic evaluation and should be tested prior to any in-depth assessment to verify the authenticity of the samples. The fruit's color, shape, odor, and taste were found to be the most important characteristics for identifying the plant in a macroscopic study. The macroscopic observations for *P. domestica* revealed the fruit to have a blackish-brown color and an oval shape. The size of the fruit ranges from 6–8 cm in diameter and the surface is smooth (Figure 1). The odor of the fruit is delicious, and the taste is sweet to acidic.

**Figure 1.** The macroscopy observation for *P. domestica* fruit sample.

### 3.2. Microscopic Observations

The transverse section of the fruit revealed an abundance of the rectangular-shaped parenchymatous cells in its ground tissue, thin-walled epidermis devoid of cuticle and any kind of excrescences, and uniformly distributed coloring matter possessing vascular bundles of xylem and phloem. The presence of crystals or any other orgastic content was not clear in the section (Figure 2). These specific characteristics are useful for the standardization of the fruit and may be used for the preparation of plant monographs in order to reduce the possibilities of adulteration.

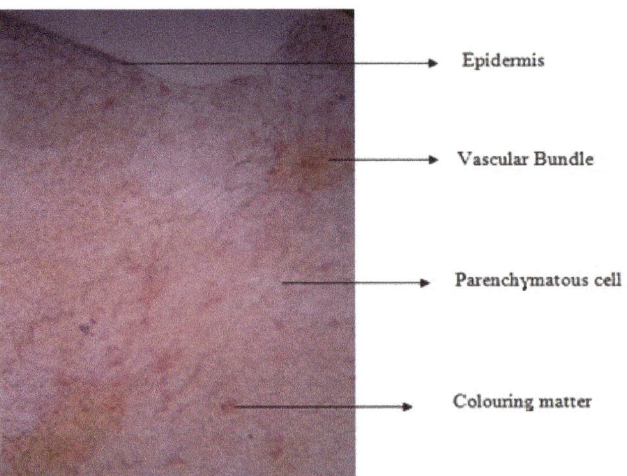



**Figure 2.** The microscopy observation for *P. domestica* fruit sample.

### 3.3. Physiochemical Analysis

The results for various physiochemical tests are presented in Table 1. The assessment of the moisture content is an essential parameter for detecting inappropriate storage and handling of a sample. Herein, the moisture content of the drug as determined with the help of loss on drying (LOD) was revealed to be within the limits specified. The total ash is required for analysis of the purity of samples, i.e., the presence or absence of foreign inorganic matter (silica, metallic salts, etc.). It is a well-known fact that the total ash value may not be enough to determine the quality of a sample or herbal drugs, as the plant materials generally have major amounts of physiological ash (calcium oxalate and earthy matters in particular). Hence, acid insoluble ash value is more appropriate to determine the quality of herbal drugs in cases in which the evaluation for silica and earthy matter is desirable. The water-soluble ash consists of a water-soluble part of the total ash used to determine the amount of inorganic material observed in the herbal drugs [18–20]. A previous physiochemical study of this plant was compared and found to have very satisfactory results, but the previous study was on seeds of the plant, whereas the present study is on fruit [21].

**Table 1.** Summary of physicochemical parameters of *P. domestica* fruit ($n = 5$).

| Parameters | % *w/w* (Mean ± SD) |
| --- | --- |
| LOD | 15.46 ± 2.24% |
| Moisture content | 13.27 ± 1.75% |
| Ash value | |
| Total ash | 3.66 ± 0.257% |
| Acid insoluble ash | 0.36 ± 0.082% |
| Water-soluble ash | 2.83 ± 0.817% |
| **Successive extraction values** | |
| Petroleum ether | 1.50 ± 0.13% |
| Chloroform | 1.8 ± 0.35% |
| Methanol | 15.21 ± 2.43% |
| Water:alcohol (1:1; *v/v*) | 24.71 ± 4.94% |
| Water | 20.80 ± 4.41% |

The amount of an extract (extractive yield) in a particular solvent is mostly an approximation of the number of specific compounds that the drug contains. The drug should be extracted using a variety of solvents in order of their increasing polarity to obtain reliable and accurate values. Generally, petroleum ether, chloroform, methanol, water:alcohol (1:1; *v/v*), and water extractives are taken into account for finalizing the standards of a drug. The petroleum ether extract contains fixed oils, resins, and volatile materials. Although heating the extract (105 °C until constant weight) may evaporate the volatile oils, resins, coloring matters, and fixed oil still remain. Though alcohol solvents may dissolve almost all active compounds, they are usually used for analyzing the extractive index of the samples containing alkaloids, glycosides, resin, etc. Water is used for drug samples containing aqueous soluble materials as their major compounds.

The results for physicochemical analysis (LOD, ash values, and extractive values) were found within the limits and comparable with pharmacopoeial standards.

### 3.4. Phytochemical Analysis

Preliminary phytochemical screening (color reactions) of *P. domestica* fruit extract revealed numerous phytoconstituents classes, including alkaloids, terpenes, carbohydrate, phenolic compounds, flavonoids, and glycosides (Table 2).

**Table 2.** Phytochemical tests for detection of chemical classes in *P. domestica* fruit.

| S. No. | Phytochemical Tests | Chloroform Extract | Alcoholic Extract | Aqueous Extract |
| --- | --- | --- | --- | --- |
| 1 | Alkaloid | + | + | + |
| 2 | Sterols | + | + | + |
| 3 | Carbohydrate | − | + | + |
| 4 | Phenolic compound | + | + | + |
| 5 | Flavonoid | + | + | + |
| 6 | Amino acids | − | + | + |
| 7 | Saponin | − | + | + |
| 8 | Mucilage | − | − | − |
| 9 | Glycoside | − | + | + |
| 10 | Terpenes | + | + | − |

+, present, −, absent.

### 3.5. Total Phenolic Content

In the plant kingdom, phenolics are the most common secondary metabolite. All of these different groups of constituents have received considerable attention as potential natural antioxidants because of their capability to act both as effective radical scavengers and metal chelators. Scientific evidence confirms that the antioxidant activity of phenol is attributed primarily to its redox properties, singlet oxygen quenchers, and hydrogen donor ability [22]. The total phenolic content in *P. domestica* extract was determined using the modified Folin–Ciocalteu method and found to be 1.98 ± 0.263% *w/w*.

### 3.6. Total Flavonoid Content

Similar to phenolics, flavonoids also possess significant antioxidant activity with considerable effects on human nourishment and health. Flavonoids work by either scavenging or chelating free radicals in the body [23]. The total flavonoid content in extract samples was assessed using the aluminum chloride colorimetric method, and the content was found to be 1.18 ± 0.484% *w/w* for *P. domestica* fruit.

### 3.7. Heavy Metals Determination via AAS

Due to their toxicity, persistence, and bioaccumulative nature, heavy metal contamination of traditional medicines and herbal samples is still a serious concern [24]. Heavy metals are well known for their side effects on several organs of the human body. For instance, continuing contact with lead may cause disturbance in the functioning of the nervous system, in addition to affecting kidney clearance [25]. The major sources of contamination with heavy metals such as cadmium, mercury, lead, and arsenic may be attributed to a contaminated environment during harvest or growth. The detection of contaminants (Pb, Cd, Hg, Ar) for *P. domestica* fruit sample in the linearity range of (0.5000–1.5000 ppm) revealed no resultant spectral peaks for any of the contaminants, as observed with AAS spectra (Table 3).

**Table 3.** Heavy metal analysis of *P. domestica* fruit ($n = 5$).

|  | Mean ± SD (ppm) | Limit (Safe Up to) (ppm) |
|---|---|---|
| Lead | 0.56301 ± 0.0089 | 10 |
| Cadmium | 0.00453 ± 0.0002 | 0.30 |
| Mercury | 0.441 ± 0.0246 | 0.50 |
| Arsenic | 1.182 ± 0.0203 | 3.0 |

### 3.8. Aflatoxins Determination via HPLC

Aspergillus genus fungi produce aflatoxins, which are mycotoxins that can grow on a variety of food materials, spices, and herbal drugs. The main classification of aflatoxins includes $B_1$, $B_2$, $G_1$, and $G_2$; in this classification, $B_1$ and $G_1$ aflatoxins are considered more toxic than $B_2$ and $G_2$ because of the presence of extra double bond, which leads to the making of an electrophilic reactive epoxide in hepatic metabolism. Acute structural and functional injury to essential organs of the human body can occur because of this mechanism in action. It was considered that aflatoxin B1 can be a risk factor in the etiology of hepatocellular cancer in humans [26,27]. Environmental conditions such as temperature and high moisture during the cultivation and storage of herbal drugs support the growth and contamination of aflatoxins by the aspergillus genus. Herein, aflatoxins were analyzed by the HPLC method, and the results obtained revealed a lack of any such aflatoxins in the *P. domestica* fruit sample.

### 3.9. Pesticides Determination via GC–MS

The ever-increasing demands for herbal drugs urge cultivation of the medicinal plants on a larger scale where the application of pesticides is also witnessed at an excessive level. Particular attention has been given to the impurity of organochlorine pesticides (OCPs) due to their toxicity and perseverance in the atmosphere and contamination by common pesticides [28,29]. The pesticides were evaluated by the GC–MS method, which revealed a pesticide-free sample of *P. domestica*. The pesticides analyzed in this study are listed in Table 4.

**Table 4.** Different types of pesticides screened by AOAC method in selected fruit.

| S. No. | Pesticide | Test Method |
|---|---|---|
| 1 | α-BHC | AOAC970.52/EPA525.5 |
| 2 | β-BHC | AOAC970.52/EPA525.5 |
| 3 | γ-BHC(Lindanee) | AOAC970.52/EPA525.5 |
| 4 | δ-BHC | AOAC970.52/EPA525.5 |
| 5 | Heptachlor | AOAC970.52/EPA525.5 |
| 6 | Heptachlor_Epoxide | AOAC970.52/EPA525.5 |
| 7 | α-Chlordane | AOAC970.52/EPA525.5 |
| 8 | α-Endoulfan | AOAC970.52/EPA525.5 |
| 9 | β-Chlordane | AOAC970.52/EPA525.5 |
| 10 | Endrin | AOAC970.52/EPA525.5 |
| 11 | Total DDE | AOAC970.52/EPA525.5 |
| 12 | Total DDD | AOAC970.52/EPA525.5 |
| 13 | Total DDT | AOAC970.52/EPA525.5 |
| 14 | β-Endoulfan | AOAC970.52/EPA525.5 |
| 15 | Endrin_Aldehyde | AOAC970.52/EPA525.5 |

**Table 4.** *Cont.*

| S. No. | Pesticide | Test Method |
|---|---|---|
| 16 | Endoulfan_sulfate | AOAC970.52/EPA525.5 |
| 17 | Aldrin | AOAC970.52/EPA525.5 |
| 18 | Endrin_Ketone | AOAC970.52/EPA525.5 |
| 19 | Methoxychlor | AOAC970.52/EPA525.5 |
| 20 | Dieldrin | AOAC970.52/EPA525.5 |
| 21 | Alachlor | AOAC970.52/EPA525.5 |
| 22 | Butachlor | AOAC970.52/EPA525.5 |
| 23 | Monochlorphos | AOAC970.52/EPA525.5 |
| 24 | Phorate | AOAC970.52/EPA525.5 |
| 25 | Mevinphos | AOAC970.52/EPA525.5 |
| 26 | Dimethoate | AOAC970.52/EPA525.5 |
| 27 | Malathion | AOAC970.52/EPA525.5 |
| 28 | Methyl-parathion | AOAC970.52/EPA525.5 |
| 29 | Chlorpyrifos | AOAC970.52/EPA525.5 |
| 30 | Ethion | AOAC970.52/EPA525.5 |
| 31 | Atrazine | AOAC970.52/EPA525.5 |
| 32 | Simazine | AOAC970.52/EPA525.5 |
| 33 | Diazinone | AOAC970.52/EPA525.5 |
| 34 | Phosphamidon | AOAC970.52/EPA525.5 |
| 35 | Fenitrothion | AOAC970.52/EPA525.5 |
| 36 | Fenthion | AOAC970.52/EPA525.5 |
| 37 | Phosalone | AOAC970.52/EPA525.5 |
| 38 | Quinaphos | AOAC970.52/EPA525.5 |
| 40 | Malaoxon | AOAC970.52/EPA525.5 |
| 41 | Dichlorvos | AOAC970.52/EPA525.5 |
| 42 | 2,4-D | AOAC970.52/EPA525.5 |

### 3.10. HPTLC Finger Printing

The technique of HPTLC is considered a mainstream approach to determine the quality of a sample with the help of fingerprint of plant standard chemicals. HPTLC has a high degree of sensitivity, which enables the detection of a variety of chemicals in a single run. The main objective of HPTLC estimation of *P. domestica* fruit was to develop an exceptional HPTLC chemical drugs pattern, representative of the whole chemical profile present in the fruit sample. A number of mobile phases were tried through the hit and trial technique for various solvent–extract ratios; acceptable separation of the compound was observed in the solvent system of toluene:ethyl acetate:formic acid (5:4:0.5; *v/v/v*), and butanol:acetic acid:water (8:2:2; *v/v/v*) for chloroform, methanol, aqueous:alcohol (1:1; *v/v*), and aqueous extract. The samples were applied, and a chromatogram was established in corresponding solvents and scanned at 366 nm. The results for solvent systems and peaks are presented in Table 5 and Figure 3.

**Table 5.** HPTLC fingerprint data of different extracts of *P. domestica* fruit.

| S. No. | Sample | Solvent System | No. of Peaks and $R_f$ Values |
| --- | --- | --- | --- |
| 1 | Chloroform extract | Toluene:Ethyl acetate:formic acid (5:4:0.5; *v/v/v*) | **(06)**; 0.05, 0.11, 0.29, 0.48, 0.68, 0.75 |
| 2 | Methanolic extract | Toluene:Ethyl acetate:formic acid (5:4:0.5; *v/v/v*) | **(08)**; 0.01, 0.20, 0.53, 0.56, 0.62, 0.66, 0.83, 0.92 |
| 3 | Aqueous:alcohol extract (1:1; *v/v*) | Toluene:Ethyl acetate:formic acid (5:4:0.5; *v/v/v*) | **(10)**; 0.01, 0.11, 0.19, 0.35, 0.37, 0.57, 0.65, 0.67, 0.83, 0.93 |
| 4 | Aqueous extract | Butanol:acetic acid:water(8:2:2; *v/v/v*) | **(11)**; 0.01, 0.15, 0.20, 0.45, 0.54, 0.55, 0.57, 0.63, 0.67, 0.84, 0.93 |

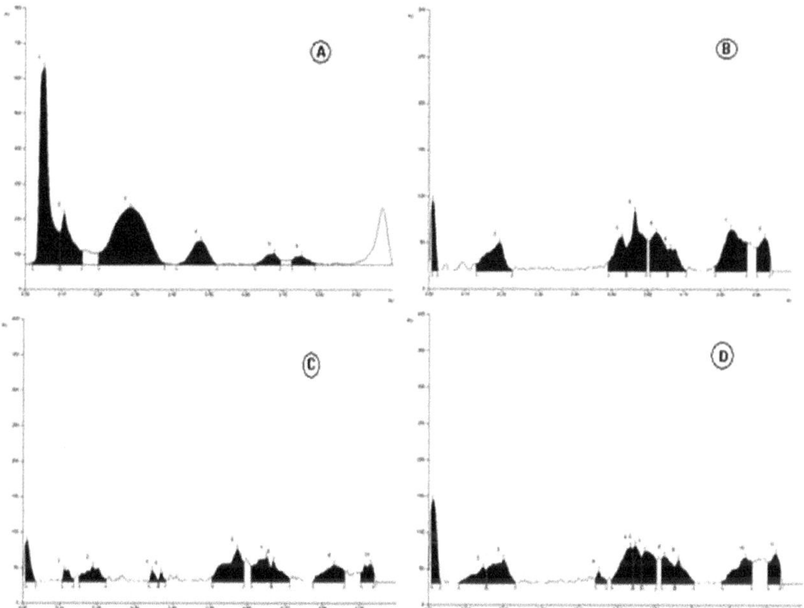

**Figure 3.** HPTLC fingerprint profile of chloroform (**A**), methanol (**B**), aqueous: alcohol (1:1 *v/v*) (**C**) and aqueous (**D**) of *P. domestica* fruit.

### 3.11. Microbial Load for the Fruit Sample

Generally, herbal drugs contain a variety of soil-borne microorganisms and molds. In order to ensure the safety of samples, the bioburden level is determined according to the procedure recommended by the WHO. The microbial profile for *P. domestica* was found within acceptable limits. The total microbial, mold, and yeast plate counts were 40 CFU/mL ($\leq$10 CFU/mL as per the WHO guidelines). Furthermore, the pathogenic bacteria (*E. coli*, *Salmonella*, *Pseudomonas*, and *Staphylococcus*) were not found in the sample. The plates with microbe growth are shown in Figure 4.

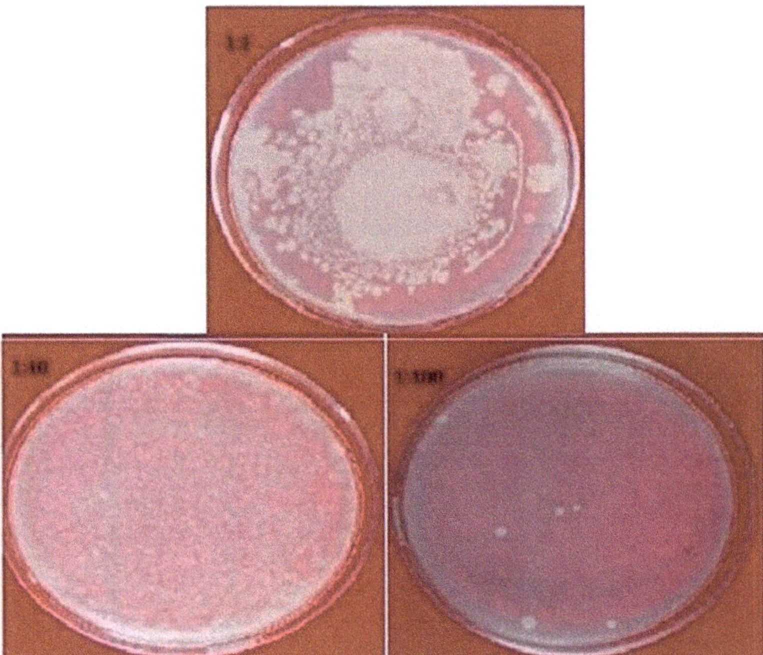

**Figure 4.** The plates representing the growth of microbe found in *P. domestica* fruit sample.

### 3.12. HPLC/DAD–DPPH Method for In Vitro Antioxidant Activity

Antioxidant activity in complex mixtures, such as herbal extracts, can be quickly assessed using HPLC–DPPH method [30]. The antioxidant potential for various extracts of the *P. domestica* fruit sample was determined by HPLC using a fluorescent detector. The half inhibition concentration ($IC_{50}$) of *P. domestica* fruit extract was $34.28 \pm 2.08$ µg/mL, while the $IC_{50}$ value for ascorbic acid was $16.30 \pm 1.32$ µg/mL (Figure 5). However, the results show that the developed method can be used to quickly screen for antioxidant compounds or more precisely radical scavenging activity of natural compounds. A previous antioxidant study of this plant was compared and found to have very satisfactory results, but the previous study was on seeds of the plant, whereas the present study is on fruit [31].

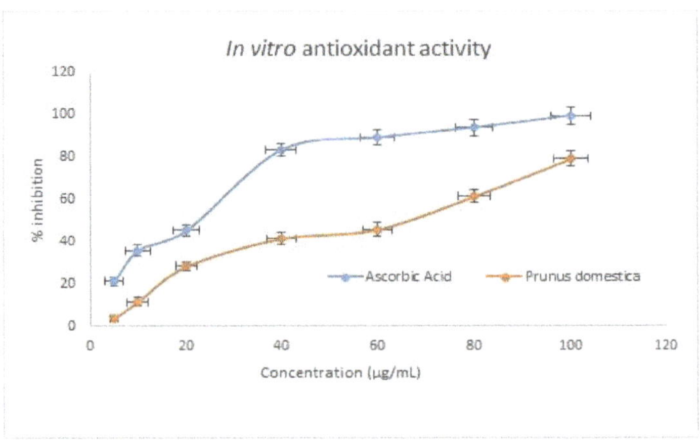


**Figure 5.** In vitro antioxidant activity.

## 4. Conclusions

The results of physicochemical analysis, phytochemical screening, heavy metal detection, pesticides level, and in vitro antioxidant activity were useful in establishing the quality, safety, and efficacy of *P. domestica* fruit for its use as a potential drug candidate. The quality evaluation study herein may be a useful tool for the identification and standardization of the plant in terms of quality control of raw materials used in the nutraceutical or herbal formulation in industries.

**Author Contributions:** M.A.; Methodology, writing, R.A.; review and editing, A.Z.; review and editing, W.A.; visualization, M.S.; visualization, M.K.; formal analysis, M.M.G.; resources, S.A. (Sultan Alshehri); visualization, resources, S.W.; visualization, S.A. (Sayeed Ahmad); validation, formal analysis, M.M.; project administration, validation and formal analysis. All authors have read and agreed to the published version of the manuscript.

**Funding:** This research received no external funding.

**Institutional Review Board Statement:** Not applicable.

**Informed Consent Statement:** Not applicable.

**Data Availability Statement:** Not applicable.

**Conflicts of Interest:** The authors declare no conflict of interest.

# References


1. Yates, A. *Yates Garden Guide*; Harper Collins Australia: Pymble, Australia, 2002.
2. Amir, M.; Ahmad, N.; Sarafroz, M.; Ahmad, S.; Mujeeb, M. Pharmacognostical, Physicochemical Standardization and In vitro Antioxidant Activity of *Punica granatum* Linn fruit. *Pharmacogn. J.* **2019**, *11*, 73–79. [CrossRef]
3. Sethi, P.-D. *High Performance Thin Layer Chromatography: Quantitative Analysis of Pharmaceutical Formulations*; CBS Publishers and Distributers: New Delhi, India, 1996; pp. 10–60.
4. *Quality Control Method for Medicinal Plant Materials*; WHO: Geneva, Switzerland, 1989; pp. 1–15.
5. Gupta, V.-K. *Wealth of India*; CSIR: New Delhi, India, 2003; Volume 4, pp. 406–409.
6. Ramming, D.-W.; Cociu, V. Plum (*Prunus*). *Genet. Resour. Temp. Fruit Nut Crops* **1991**, *290*, 235–287.
7. Chiej, R. *MacDonald Encyclopaedia of Medicinal Plants*; MacDonald: London, UK, 1984; p. 448.
8. Kayano, S.; Kikuzaki, H.; Ikami, T.; Suzuki, T.; Mitani, T.; Nakatani, N. A New Bipyrrole and Some Phenolic Constituents in Prunes (*Prunus domestica* L.) and Their Oxygen Radical Absorbance Capacity (ORAC). *Biosci. Biotechnol. Biochem.* **2004**, *68*, 942–944. [CrossRef] [PubMed]
9. Baranowski, R.; Kabut, J.; Baranowska, I. Analysis of mixture of catechins, flavones, flavanones, flavonols, and anthocyanidins by RP-HPLC. *Anal. Lett.* **2004**, *37*, 157–165. [CrossRef]
10. Tsuji, M.; Harakawa, M.; Komiyama, Y. High performance liquid chromatographic analysis of quinic acid malic acid and shikimic acid in plum fruit using a cation exchange resin. *J. Jpn. Soc. Food Sci. Technol. (Nippon. Shokuhin Kogyo Gakkaishi)* **1985**, *32*, 661–663.
11. Brain, K.-R.; Turner, T.-D. *The Practical Evaluation of Phytopharmaceuticals*; Wright-Scientechnica: Bristol, UK, 1975.
12. Khandelwal, K.-R. *Practical Book of Pharmacognosy*; Nirali Prakashan: Pune, India, 2005.
13. Amir, M.; Mujeeb, M.; Khan, A.; Ashraf, K.; Sharma, D.; Aqil, M. Phytochemical analysis and in vitro antioxidant activity of *Uncaria gambir*. *Int. J. Green Pharm.* **2012**, *6*, 67–72.
14. Horwitz, W. *Official Method of Analysis of the Association of Official Analytical Chemists*, 7th ed.; AOAC: Washington, DC, USA, 1970.
15. Ahmad, A.; Amir, M.; Alshadidi, A.-A.; Hussain, M.D.; Haq, A.; Kazi, M. Central composite design expert-supported development and validation of HPTLC method: Relevance in quantitative evaluation of protopine in *Fumaria indica*. *Saudi Pharm. J.* **2020**, *28*, 487–494. [CrossRef]
16. Rajput, S.; Tripathi, M.-K.; Tiwari, A.-K.; Dwivedi, N.; Tripathi, S.-P. Scientific evaluation of *Panchkola churna*–an Ayurvedic polyherbal drug formulation. *Indian J. Tradit. Knowl.* **2012**, *11*, 697–703.
17. Yen, G.-C.; Chen, H.Y. Antioxidant activity of various tea extracts in relation to their antimutagenicity. *J. Agric. Food Chem.* **1995**, *43*, 27–32. [CrossRef]
18. Dave, R.; Nagani, K.; Chanda, S. Pharmacognostic studies and physicochemical properties of the *Polyalthia longifolia* var. pendula leaf. *Pharmacogn. J.* **2010**, *2*, 572–576. [CrossRef]
19. Vaghasiya, Y.; Nair, R.; Chanda, S. Antibacterial and preliminary phytochemical and physico-chemical analysis of *Eucalyptus citriodora* Hk leaf. *Nat. Prod. Res.* **2008**, *22*, 754–762. [CrossRef] [PubMed]
20. Prasanth, D.S.N.B.K.; Rao, A.-S.; Prasad, Y.-R. Assessment of Pharmacognostic, Phytochemical and Physicochemical Standards of *Aralia racemosa* (L.) root. *Ind. J. Pharm. Educ. Res.* **2016**, *50*, S225–S230. [CrossRef]

21. Savic, I.; Savic Gajic, I.; Gajic, D. Physico-Chemical Properties and Oxidative Stability of Fixed Oil from Plum Seeds (*Prunus domestica* Linn.). *Biomolecules* **2020**, *10*, 294. [CrossRef]
22. Rice-Evans, C.-A.; Miller, N.-J.; Bollwell, P.-G.; Bramley, P.M.; Pridham, J.B. The relative antioxidant activities of plant-derived polyphenolic flavonoids. *Free Radic. Res.* **1995**, *22*, 375. [CrossRef] [PubMed]
23. Kessler, M.; Ubeaud, G.; Jung, L. Anti- and pro-oxidant activity of rutin and quercetin derivatives. *J. Pharm. Pharmacol.* **2003**, *55*, 131–142. [CrossRef] [PubMed]
24. Ikem, A.; Egiebo, N.-O.; Nyavor, K. Trace elements in water, fish and sediment from Tuskegee Lake, Southeastern USA. *Water Air Soil Pollut.* **2003**, *149*, 51–75. [CrossRef]
25. Samuel, J.-B.; Stanley, J.-A.; Princess, R.-A.; Shanthi, P.; Sebastian, M.-S. Gestational cadmium exposure-induced ovotoxicity delays puberty through oxidative stress and impaired steroid hormone levels. *J. Med. Toxicol.* **2011**, *7*, 195–204. [CrossRef]
26. Eaton, D.-L.; Groopman, J.-D. *The Toxicology of Aflatoxins: Human Health, Veterinary and Agricultural Significance*, 1st ed.; Academic Press: San Diego, CA, USA, 1994; p. 309.
27. McGlynn, K.; Rosvold, E.; Lustbader, E.; Hu, Y.; Clapper, M.; Zhou, T.; Wild, C.-P.; Xia, X.-L.; Baffoe-Bonnie, A.; Ofori-Adjei, D. Susceptibility to hepatocellular carcinoma is associated with genetic variation in the enzymatic detoxification of aflatoxin B1. *Proc. Natl. Acad. Sci. USA* **1995**, *92*, 2384–2387. [CrossRef]
28. Tewary, D.-K.; Kumar, V.; Shanker, A. Leaching of pesticides in herbal decoction. *Chem. Health Saf.* **2004**, *11*, 25–29. [CrossRef]
29. Sankararamakrishnan, N.; Sharma, A.-K.; Sanghi, R. Organochlorine and organophosphorus pesticide residues in ground water and surface waters of Kanpur, Uttar Pradesh, India. *Environ. Int.* **2005**, *31*, 113–120. [CrossRef]
30. Geng, D.; Chi, X.; Dong, Q.; Hu, F. Antioxidants screening in *Limonium aureum* by optimized online HPLC–DPPH assay. *Ind. Crops Prod.* **2015**, *67*, 492–497. [CrossRef]
31. Savic, I.M.; Savic Gajic, I.M. Optimization study on extraction of antioxidants from plum seeds (*Prunus domestica* L.). *Optim. Eng.* **2021**, *22*, 141–158. [CrossRef]

*plants*

# Effect of Processing on Bioactive Compounds, Antioxidant Activity, Physicochemical, and Sensory Properties of Orange Sweet Potato, Red Rice, and Their Application for Flake Products

Ignasius Radix A. P. Jati [1,*], Laurensia M. Y. D. Darmoatmodjo [1], Thomas I. P. Suseno [1], Susana Ristiarini [1] and Condro Wibowo [2]

1 Department of Food Technology, Widya Mandala Surabaya Catholic University, Jl. Dinoyo 42-44, Surabaya 60265, Indonesia; yulian@ukwms.ac.id (L.M.Y.D.D.); thomasindartoftp@gmail.com (T.I.P.S.); ristiarini@ukwms.ac.id (S.R.)
2 Department of Food Technology, Jenderal Soedirman University, Jl. Dr. Soeparno 61, Purwokerto 53123, Indonesia; condro.wibowo@unsoed.ac.id
* Correspondence: radix@ukwms.ac.id; Tel.: +62-31-5678478 (ext. 110)

**Citation:** Jati, I.R.A.P.; Darmoatmodjo, L.M.Y.D.; Suseno, T.I.P.; Ristiarini, S.; Wibowo, C. Effect of Processing on Bioactive Compounds, Antioxidant Activity, Physicochemical, and Sensory Properties of Orange Sweet Potato, Red Rice, and Their Application for Flake Products. *Plants* **2022**, *11*, 440. https://doi.org/10.3390/plants11030440

Academic Editor: Ivo Vaz de Oliveira

Received: 7 January 2022
Accepted: 1 February 2022
Published: 5 February 2022

**Publisher's Note:** MDPI stays neutral with regard to jurisdictional claims in published maps and institutional affiliations.



**Abstract:** Orange sweet potato (OSP) and red rice (RR) are rich sources of health benefit-associated substances and can be conventionally cooked or developed into food products. This research approach was to closely monitor the changes of bioactive compounds and their ability as antioxidants from the native form to the food products which are ready to be consumed. Moreover, this research explored the individual carotenoids and tocopherols of raw and cooked OSP and RR and their developed flake products, and also investigated their antioxidant activity, physicochemical properties, and sensory properties. Simultaneous identification using the liquid chromatographic method showed that OSP, RR, and their flake products have significant amounts ($\mu g/g$) of $\beta$-carotene (278.58–48.83), $\alpha$-carotene (19.57–15.66), $\beta$-cryptoxanthin (4.83–2.97), $\alpha$-tocopherol (57.65–18.31), and also $\gamma$-tocopherol (40.11–12.15). Different responses were observed on the bioactive compound and antioxidant activity affected by heating process. Meanwhile, OSP and RR can be combined to form promising flake products, as shown from the physicochemical analysis such as moisture (5.71–4.25%) and dietary fiber (13.86–9.47%) contents, water absorption index (1.69–1.06), fracturability (8.48–2.27), crispness (3.9–1.5), and color. Those quality parameters were affected by the proportions of OSP and RR in the flake products. Moreover, the preference scores (n = 120 panelists) for the flakes ranged from slightly liked to indifferent. It can be concluded that OSP and RR are potential sources of bioactive compounds which could act as antioxidants and could be developed into flake products that meet the dietary and sensory needs of consumers.

**Keywords:** orange sweet potato; red rice; flakes; bioactive compound; antioxidant activity; physicochemical; sensory properties

## 1. Introduction

Modern food trends and lifestyle changes have strongly influenced the dietary habits of society. The demands for ready-to-eat and simple-to-prepare foods are increasing rapidly, providing an excellent opportunity for food industries to play a significant role in supplying such food products. Flakes, one of the most popular foods made from cereals, typically oat, corn, and barley, are commonly served for breakfast with milk in Europe and the USA, and their global appeal is gaining traction. In addition, healthy eating has become a new trend in modern culture, with consumers increasingly opting for healthy food options. Secondary metabolites found in plants have been shown to decrease the risks of degenerative diseases such as coronary heart disease, diabetes, cancer, and stroke [1–4]. Thus, innovative functional food products rich in bioactive compounds are needed to promote a healthy diet and reduce disease risks. Asia has the most rapidly growing food product market, so using local ingredients can help open a new market for flakes and decrease the reliance on imported foods such as oat and barley. Commodities that can potentially be developed into flakes are red rice (RR) and orange sweet potato (OSP).

RR (*Oryza nivara* L.) is a variety of rice with red pericarp caused by anthocyanins in the aleurone layer. It is a rich source of anthocyanins such as cyanidin 3-O-glucoside and peonidin 3-O-glucoside [2]. Anthocyanins have also been reported to inhibit plaque formation [3] and to exhibit hypocholesterolemic [4] and anticancer effects [5] in RR. The health benefits of RR have also been linked to bioactive compounds such as tocopherol and tocotrienols [6,7] and dietary fiber [8]. RR also contains higher essential minerals than white rice, including iron, zinc, and vitamins, which are especially important for babies and toddlers, [9]. Despite the various health benefits of RR, its consumption remains low. It is generally considered inferior to white rice due to the hard texture and unpleasant aroma when cooked. Traditionally, RR is consumed steamed or boiled, and its sensory properties are inferior to white rice. The most popular RR-based product is RR baby porridge, but there are reports on the development of RR-based products such as pasta [10], noodles [11], flakes [12], rice milk [13], and fermented beverages [14]. However, these have not been scaled up or commercially established.

OSP (*Ipomoea batatas*) is one variety of sweet potato with a bright orange flesh color caused by carotenoids, of which high amounts of β-carotene, α-carotene, and β-cryptoxanthin have been reported [15–17]. In many countries, OSP has been used to eradicate vitamin A deficiency due to its high content of beta-carotene, a pro-vitamin A carotenoid. Sweet potato was extensively promoted in Africa and some Asian countries with remarkable results [18–20]. However, there were difficulties in ensuring its sustainability due to the monotonous method of preparation, mainly boiled and baked, even though the essential nutritional components were reportedly retained after processing [21]. Processing OSP could decrease the beta-carotene, but not below the recommended dietary level [22]. Moreover, food prepared from OSP by the traditional method was not attractive to children, who were the main target of the vitamin A intake enhancement. Therefore, innovative food products need to be developed to promote the consumption of RR and OSP. Numerous research has been published to explore the potency of various plant sources as functional foods. However, the approach mainly investigates the plant materials in the native or raw form. In contrast, the consumer will generally consume after the materials undergo transformations which could affect the characteristics of the products [23]. Moreover, besides the processing, the bioactive compounds and antioxidant activity will be further affected by the in vivo digestion and the intestinal absorption rate of the body metabolism before providing bioavailable compounds that can be used [24]. This research approach was to closely monitor the changes of the bioactive compound and antioxidant activity from the raw materials to the ready to be consumed food products, and aimed to investigate the bioactive compounds and antioxidant activity of raw and cooked RR and OSP and the physicochemical and sensory properties of their developed flake products.

## 2. Materials and Methods

### 2.1. Plant Materials and Chemicals

A local variety of RR (*Oryza nivara* L.), "Cempo abang", and OSP (*Ipomoea batatas* L.), "Mendut", were collected from farmers in Yogyakarta province, Indonesia. Chemicals used for analysis, including distilled water, Folin–Ciocalteu, 1,1-diphenyl-2-picrylhydrazyl (DPPH), gallic acid, butylated hydroxyl toluene (BHT), enzymes (thermamyl, pancreatin, pepsin), riboflavin, methionine, and nitroblue tetrazolium (NBT), were purchased from Sigma Chemical. Methanol, Whatman 40 filter paper, n-hexane, NaOH, HCl, ethanol, and phosphate buffer (pH 6) were purchased from Merck, Germany. Carotenoid standards, including β- and α-carotene, β-cryptoxanthin, lycopene, and lutein, and α- and γ-tocopherol standards were obtained from Sigma-Aldrich.

*2.2. Sample Preparation*

2.2.1. Raw Samples

The RR samples were washed, drained, and blended (Philips food processor). The OSP samples were chopped into small pieces. All samples were then freeze-dried, refined, and sieved (30 mesh). Finally, the powdered samples were placed in dark bottles and stored in a refrigerator (4 °C) for further usage.

2.2.2. Boiled Samples

RR (75 g) was cooked with 135 g of tap water (1:1.8; *w/w*) in a rice cooker (Panasonic) for approximately 45 min. After, the cooked rice was cooled for 10 min. Meanwhile, 150 g of OSP was boiled in an aluminum pot using tap water for 20 min, cooled for 10 min, and then mashed. Both samples were freeze-dried, refined, sieved (30 mesh), and stored (4 °C) in a refrigerator.

2.2.3. Flakes Production

RR was placed in a cabinet dryer (60 °C) for 1 h. The OSP was peeled, sliced and placed in a cabinet dryer (60 °C) for 6 h. The dried OSP and RR were mashed using a blender. The flour was passed through an 80 mesh. Flakes were produced using six different proportions of OSP and RR, namely 100:0, 80:20, 60:40, 40:60, 20:80, and 0:100. Salt (3% *w/w*), sugar (30% *w/w*), and water (150% *w/w*) were mixed in as additional ingredients. The mixture was heated at 75 °C for 1 min and pressed at 170 °C for 1 min using a customized flake pressing tool. The pressed flakes were cut at 2 × 2 cm and dried using an oven at 125 °C for 5 min.

*2.3. Bioactive Compounds and Antioxidant Activity Analysis*

2.3.1. Methanolic Extract of Samples

The extraction of samples (raw, boiled, and flakes) was performed according to a previously published procedure [25]. Briefly, 1 g of sample was weighed, ground, placed in a centrifuge tube, and then extracted with 10 mL of 1% methanol–HCl solution. The mixture was then vortexed for 15 min, centrifuged at 5000 rpm for 15 min, filtered (Whatman No. 40), and then used for antioxidant activity analysis, performed in triplicate.

2.3.2. Phenolic Content

The total phenolic content was determined using the Folin–Ciocalteu method by Singleton and Rossi as described in other published work [26]. In brief, 0.1 mL of extract was mixed with 0.5 mL 1:1 Folin–Ciocalteu reagent and distilled water. After 10 min, 4.5 mL of 2% sodium carbonate ($Na_2CO_3$) was added, and the mixture was then vortexed and kept in the dark for 1 h. The blue complex formed was measured using a spectrophotometer at 765 nm. Methanol and gallic acid were used as the blank and standard, respectively. The results were calculated as milligram gallic acid equivalents (GAE)/100 g dry weight.

2.3.3. Anthocyanin Content

The total anthocyanin in RR and flake samples was determined spectrophotometrically using the pH differential method [27]. In brief, 1 mL of extract was diluted in pH 1.0 and pH 4.5 buffers. The absorbance was measured at 510 and 710 nm. The final absorbance was calculated using the formula:

$$A = [(A513\text{-}A700)pH\ 1.0 - (A513\text{-}A700)pH\ 4.5)] \tag{1}$$

The calculated absorbance was then used to calculate the total grams of anthocyanins per 100 g dry weight, with a molar extinction coefficient of 26,900 and a molecular weight of 445.

### 2.3.4. Carotenoid and Tocopherol Analysis

The carotenoids and tocopherols in the raw and boiled samples of RR and OSP were determined simultaneously using High-Performance Liquid Chromatography (HPLC) based on previously published research [24]. In brief, 0.3 g of finely ground samples were extracted with 0.5 mL of 70% ethanol and 0.4 mL of n-hexane under yellow lights. In addition, 0.4 mL β-Apo-8′carotenal-O-methyloxim and 0.4 mL α-, γ-tocopherol were used as an internal standard for carotenoids and tocopherols, respectively. The mixture was shaken for 20 min, centrifuged at 5000 rpm, 4 °C for 20 min. Next, the upper layer of extract containing a hexane fraction was collected using a micropipette. The extraction was repeated four times using only n-hexane as a solvent. Finally, the hexane fractions were pooled and completely dried using nitrogen gas.

Before injection, the extracts were mixed with 200 μL of ethanol containing 30 μg/mL BHT, then 20 μL of the mixture was injected into the HPLC (Varian Pro Star 410, Spark, The Netherlands). A mixture of 82% acetonitrile, 15% dioxan, 3% methanol, 0.1 M ammonium acetate, and 0.1% triethylamine was assigned as the mobile phase and was pumped at a rate of 1.6 mL/min. The solvent was pre-mixed to avoid dependency on reproducible mixing by the pump. For separation, a C18 Spherisorb ODS 2 column (3 μm, 250 × 4.6 mm) was applied. In addition, a UV Vis detector at 450 nm and a Scanning Fluorescence detector using an excitation wavelength of 295 nm and an emission wavelength of 328 nm were used to monitor the carotenoids and tocopherols, respectively. Five repetitions were performed for the HPLC analysis.

### 2.3.5. DPPH Radical Scavenging Activity

The radical scavenging activity of the extract was examined by the DPPH method [25]. In brief, 1 mL of extract was mixed with 2 mL of 0.2 M DPPH and 2 mL of methanol in centrifuge tubes, vortexed, and kept in the dark for 1 h. The absorbance was measured spectrophotometrically at 517 nm. As a control, 150 ppm BHT solution was used. The DPPH radical scavenging activity of the extract was expressed as a percentage calculated as follows: % radical scavenging capacity = ((Absorbance of control − Absorbance of sample)/Absorbance of control) × 100%

### 2.3.6. FRAP Assay

The ferric reducing antioxidant power (FRAP) was examined based on a previously published report [28]. In brief, a mixture of 60 μL extract, 180 μL distilled water, and 1.8 mL FRAP reagent was vortexed and incubated at 37 °C for 30 min. The spectrophotometer was used to read the absorbance of the mixture at 593 nm. A standard curve was prepared with Fe [II] (FeSO₄.7H₂O, 100–2000 mM) to calculate the reducing power. The result was expressed as mmol Fe[II]/g. In addition, methanol was used for the reagent blank.

### 2.3.7. Superoxide Radical Scavenging Capacity

A previous report [29] was followed to examine the Superoxide radical scavenging capacity. Firstly, a reagent containing riboflavin, methionine, and NBT in 0.05 M phosphate buffer pH 7.8 was prepared. Then, 100 μL of the extract was mixed with 4.9 mL of reagent and illuminated (20 W fluorescent lamp) at 25 °C for 25 min. The absorbance was measured at 560 nm.

### *2.4. Physicochemical Properties of Flake Products*
### 2.4.1. Moisture and Dietary Fiber Contents

The moisture content was measured thermogravimetrically [30]. In brief, 1 g of each sample was dried at 105 ± 0.2 °C to establish a constant mass. The analysis was performed in triplicate.

### 2.4.2. Water Absorption Index

The water absorption index was examined according to previously published method [31]. In brief, 5 g of flakes was placed in a 100 mL beaker, 30 mL of water at 30 °C was added. After 10 s of immersion, the flakes were dried. The water absorption index was calculated using the formula: WAI = (wf -wi)/wi, where wi and wf are the initial and final weight of the sample, respectively.

### 2.4.3. Fracturability, Crispness, and Color Profiles

The fracturability and crispness of flakes were measured using TA-XT Plus Texture Analyzer (Stable Micro Systems, Surrey, UK) [32]. The probe used was a $\frac{1}{4}$ inch spherical stainless-steel probe (P0.25S). The sample was placed on the sample holder, and then the probe was moved down to press the sample. The results were obtained in the form of a graph (force vs. time) (the graph is not shown). The value of the $y$-axis at the graph's highest point is the maximum force value that can be held by the sample, called the value of fracturability. Crispness can be measured through changes in displacement distance during a drastic decline in the graph pattern from the highest peak to the next peak point. Meanwhile, the color profiles of flakes were measured using color reader Konica Minolta CR-10 (Konica Minolta, Osaka, Japan). The results were expressed as Lightness (L*), redness (a*), yellowness (b*), °hue (°h), and Chroma (C).

### 2.5. Sensory Analysis

The sensory evaluation was conducted by 120 untrained panelists to determine the level of consumer preference for the flakes with various proportions of OSP and RR. The parameters tested were preferences for color, taste, crispness of flakes, and mouthfeel when served with milk. The Hedonic Scale Scoring method (preference test) with a scale ranging from 1 (strongly disliked) to 7 (strongly liked) was used for the sensory test.

Samples of flakes for the color preference test were prepared in open white plastic containers. Panelists were asked first to assess aspects of flakes' taste, color, and crispness before serving with milk. The crispness was evaluated based on the panelist's preference for the sound of flakes during biting. For the mouthfeel test, 5 g of flakes were prepared in a small plastic container. Panelists were instructed to pour 10 mL of milk into the container and wait for 1 min. Then, they were asked to assess the mouthfeel of the flakes based on preference level by filling the questionnaire sheet provided.

### 2.6. Statistical Analysis

The data were statistically analyzed using ANOVA ($\alpha$ = 5%) followed by Duncan's Multiple Range Test (DMRT) on SPSS software version 19. Spider web chart analysis using Microsoft Excel was used to determine the best proportion of OSP and RR in flakes based on the panelists' preferences.

## 3. Results

### 3.1. Bioactive Compounds of OSP, RR, and Their Flake Products

Table 1 shows the phenolic compound, anthocyanin, carotenoid, and tocopherol contents of raw and cooked OSP and RR and their flakes containing different proportions of OSP and RR. The raw RR had a higher phenolic content than OSP. As a result, the higher the proportion of RR, the higher the phenolic content of the developed flakes. It was also found that cooking decreased the phenolic content of RR and OSP by approximately 49.34% and 41.08%, respectively.

**Table 1.** Bioactive compounds of orange sweet potato (OSP), red rice (RR) and the flake products.

| | OSP | | RR | | Proportions of OSP and RR in Flakes | | | | | |
|---|---|---|---|---|---|---|---|---|---|---|
| | Raw | Cooked | Raw | Cooked | 100:0 | 80:20 | 60:40 | 40:60 | 20:80 | 0:100 |
| Phenolic (mg GAE/100 g DW) | 110.68 ± 18.3 [a] | 65.21 ± 7.3 [b] | 301.89 ± 24.86 [a] | 152.91 ± 28.92 [b] | 77.46 ± 8,28 [a] | 97.34 ± 8.79 [b] | 102.03 ± 11.65 [c] | 131.79 ± 10.93 [d] | 146.09 ± 15.64 [e] | 162.40 ± 21.54 [f] |
| Anthocyanin (mg/100 g DW) | nd | nd | 8,81 ± 0.05 [a] | 8.64 ± 0.08 [a] | nd | nd | 1.74 ± 0.06 [c] | 2.09 ± 0.05 [d] | 3.73 ± 0.07 [e] | 5.81 ± 0.04 [f] |
| $\beta$-carotene (µg/g) | 278.58 ± 31.5 [a] | 134.17 ± 17.2 [b] | 13.17 ± 2.62 [a] | 7.37 ± 0.5 [b] | 48,83 ± 3,31 [a] | 36.27 ± 3.01 [b] | 25.77 ± 3.45 [c] | 27.23 ± 2.72 [d] | 15.69 ± 2.21 [e] | 3.12 ± 0.66 [f] |
| $\alpha$-carotene (µg/g) | 19.57 ± 1.8 [a] | 23.83 ± 1.6 [b] | 5.53 ± 1.4 [a] | 11.66 ± 1.5 [b] | 15.61 ± 1.44 [a] | 11.82 ± 3.11 [b] | 5.31 ± 1.59 [c] | 2.53 ± 0.87 [d] | nd | nd |
| $\beta$-cryptoxanthin (µg/g) | 4.83 ± 0.2 [a] | 4.48 ± 0.8 [a] | 3.67 ± 2.15 [a] | 3.96 ± 1.9 [a] | 2.64 ± 0,05 [a] | 2.81 ± 0.13 [b] | 2.97 ± 0.08 | 2.78 ± 0.12 [c] | 2.77 ± 0.25 [c] | 2.81 ± 0.16 [b,c] |
| Lutein (µg/g) | 3.77 ± 0.8 [a] | 3.81 ± 0.7 [a] | 2.16 ± 0.8 [a] | 1.82 ± 0.5 [b] | nd | nd | nd | nd | nd | nd |
| $\alpha$-tocopherol (µg/g) | 13,23 ± 1.1 [a] | 15.11 ± 0,5 [b] | 34.08 ± 2.2 [a] | 57.65 ± 2.1 [b] | 4.58 ± 0.73 [a] | 7.34 ± 1.49 [b] | 10.51 ± 1.27 [c] | 12.45 ± 1.21 [c] | 16.82 ± 0.52 [e] | 18.31 ± 0.77 [f] |
| $\gamma$-tocopherol (µg/g) | 2.40 ± 0,2 [a] | 5.38 ± 0.05 [b] | 29.27 ± 2.4 [a] | 40,11 ± 1.8 [b] | nd | nd | 3.38 ± 1.22 [c] | 6.71 ± 1.19 [d] | 8.06 ± 0.98 [e] | 12.15 ± 0.73 [f] |

Data are presented as mean ± standard deviation. Different superscript letters (a–f) denote significantly different values according to Duncan's test ($p < 0.05$). Comparison was made within each category (OSP, RR, and Flakes).

Interestingly, the flakes with 100% RR showed a higher phenolic content than the cooked RR. Anthocyanin was only observed in RR. Furthermore, flakes containing 100% RR had a lower anthocyanin content than the conventionally cooked RR.

Of the carotenoids, OSP had a higher content of $\beta$-carotene, $\alpha$-carotene, $\beta$-cryptoxanthin, and lutein. $\beta$-carotene and $\beta$-cryptoxanthin were the dominant carotenoids observed in RR. The cooking process significantly decreased the content of $\beta$-carotene in OSP and RR by roughly 48% and 56%, respectively. Overall, flakes containing higher amounts of OSP showed higher carotenoid contents. The processing significantly decreased the carotenoids in the flake products when considering the raw forms and the proportions of OSP and RR, and $\beta$-carotene was the major carotenoid remaining in the flake products.

Moreover, raw RR contained higher $\alpha$-tocopherol than OSP in raw forms. Unlike the decreasing trend observed in other bioactive compounds due to processing, the tocopherol and $\alpha$-carotene contents of both RR and OSP increased after conventional cooking. On the other hand, the two-step thermal processing decreased the tocopherol content of flakes.

### 3.2. Antioxidant Activity of Raw and Cooked OSP and RR and Their Flake Products

The antioxidant activity of raw and cooked OSP and RR and their flake products were examined using DPPH, FRAP, and Superoxide radical scavenging activity methods. Figure 1a shows the DPPH scavenging activity of methanolic extract of RR, OSP, and the flake products. Boiling affected the ability of the methanolic extract to scavenge DPPH radicals. Approximately 16% and 23% decreases were observed in cooked RR and OSP, respectively. The combination of OSP and RR in the ratio of 60:40 resulted in flakes with the highest antioxidant activity (84%). The results trend indicated that the higher proportion of OSP contributed to the more robust antioxidant capacity of the extract. Furthermore, the DPPH result was in agreement with FRAP (Figure 1b) and Superoxide scavenging capacity (Figure 1c). Thus, conventional cooking and flake processing methods reduce the antioxidant activity of extracts of OSP and RR compared to their raw forms, and the right combination of OSP and RR in the flake formulation is critical for higher antioxidant activity.

### 3.3. Physicochemical Properties of OSP- and RR-Based Flake Products

The proportion of OSP and RR in the flake formulation affected the moisture content of flakes. A lower OSP proportion resulted in a lower moisture content of flakes (Table 2). The dietary fiber content increased with a higher proportion of RR. The dietary fiber content of flake products ranged from $9.47 \pm 0.01\%$ to $13.86 \pm 0.73\%$.

The water absorption index of flakes was lowest at an OSP to RR ratio of 40:60, with $0.96 \pm 0.03\%$, and was generally higher at combination ratios of 100:0, 80:20, 0:100, and 20:80. In addition, the texture characteristic of flakes was determined by the fracturability and crispness. The highest fracturability value was found in flakes made from 100% OSP ($8.48 \pm 0.09$). The reduction of the OSP proportion in flakes caused a decrease in fracturability until the proportion of 40:60 ($2.27 \pm 0.04$), beyond which the fracturability of flakes increased. A similar trend was observed in the crispness value of flakes. The lowest crispness value was detected in flakes with an OSP to RR ratio of 60:40, while higher values were obtained at ratios of 100:0, 0:100, 20:80, and 80:20.

Furthermore, the color of flakes was affected by the color of OSP and RR. The color profile of the flakes is shown in Table 3.

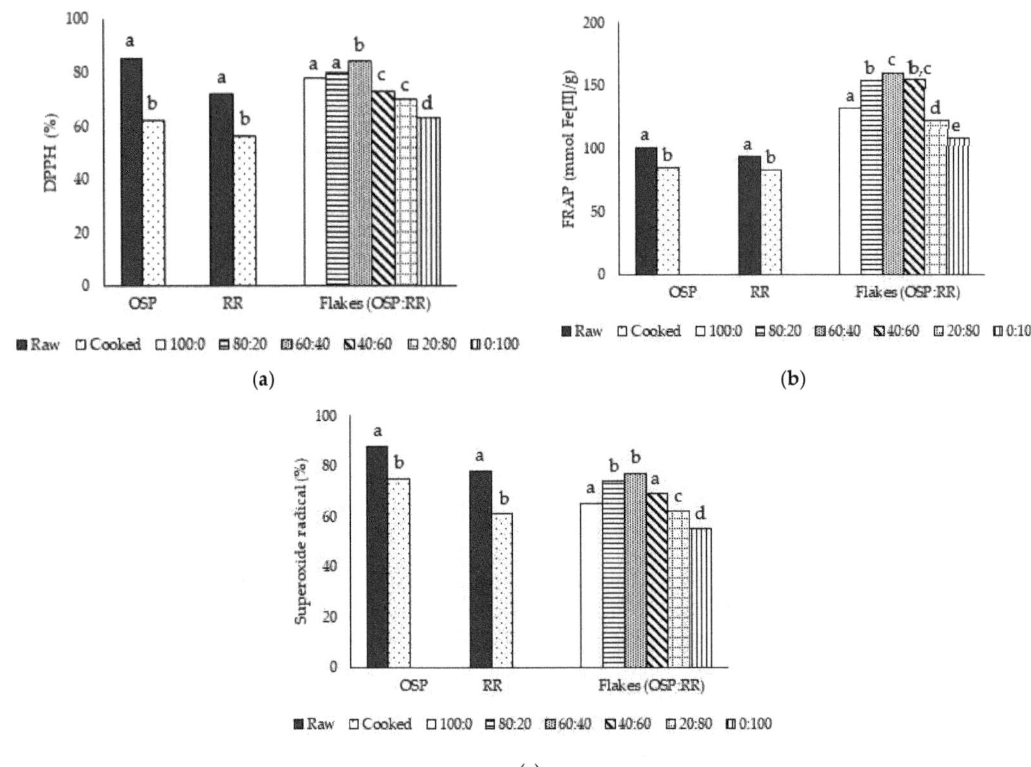

**Figure 1.** Antioxidant activity of extract determined by (**a**) DPPH, (**b**) FRAP, and (**c**) Superoxide radical assays. Different superscript letters (a–e) denote significantly different values according to Duncan's test ($p < 0.05$). Comparison was made within each category (OSP: orange sweet potato, RR: red rice, and flakes).

**Table 2.** The physicochemical properties of flakes produced from different ratios of Orange Sweet Potato (OSP), Red Rice (RR).

| | Proportions of OSP and RR in Flakes | | | | | |
|---|---|---|---|---|---|---|
| | **100:0** | **80:20** | **60:40** | **40:60** | **20:80** | **0:100** |
| Moisture content (%) | 5.71 ± 0,07 [a] | 5.31 ± 0.10 [b] | 5.09 ± 0.06 [c] | 4.87 ± 0.01 [d] | 4.43 ± 0.03 [e] | 4.25 ± 0.03 [f] |
| Dietary fiber (%) | 9.47 ± 0,01 [a] | 9.9 ± 0.02 [b] | 10.9 ± 0.05 [c] | 11.63 ± 0.34 [d] | 12.73 ± 0.26 [e] | 13.86 ± 0.73 [f] |
| Water absorption index | 1.69 ± 0,03 [a] | 1.14 ± 0.02 [b] | 1.06 ± 0.03 [c] | 0.96 ± 0.03 [d] | 1.09 ± 0.03 [b,c] | 1.12 ± 0.02 [b,c] |
| Fracturability | 8.48 ± 0,09 [a] | 5.35 ± 0.85 [b] | 3.34 ± 0.34 [c] | 2.27 ± 0.04 [d] | 3.17 ± 0.09 [e] | 4.64 ± 0.12 [f] |
| Crispness | 3.9 ± 0,03 [a] | 2.4 ± 0.02 [b] | 1.5 ± 0.02 [c] | 1.9 ± 0.03 [d] | 3.21 ± 0.05 [e] | 3.7 ± 0.03 [f] |

Data are presented as mean ± standard deviation. Different superscript letters (a–f) denote significantly different values according to Duncan's test ($p < 0.05$). Comparison was made within each row.

**Table 3.** Color profiles of flakes produced from different ratios of orange sweet potato (OSP), red rice (RR).

| | Proportions of OSP and RR in Flakes | | | | | |
|---|---|---|---|---|---|---|
| | 100:0 | 80:20 | 60:40 | 40:60 | 20:80 | 0:100 |
| L* | 44.0 ± 0.1 | 47.3 ± 0.2 | 51.5 ± 0.1 | 52.7 ± 0.4 | 51.8 ± 0.2 | 51.8 ± 0.7 |
| a* | 8.2 ± 0.3 | 8,5 ± 0.3 | 8.9 ± 0.4 | 9.4 ± 0.4 | 10.2 ± 0.6 | 10.8 ± 0.4 |
| b* | 16.5 ± 0.3 | 14.7 ± 0.2 | 13.4 ± 0.2 | 10.3 ± 0.2 | 9.0 ± 0.4 | 5.9 ± 0.4 |
| °h | 63.574 | 59.9622 | 56.4087 | 47.6158 | 41.4237 | 28.6476 |
| C | 18.47 | 16.9685 | 16.0703 | 13.898 | 13.56 | 12.3145 |

L*: Lightness; a*: redness; b*: yellowness; °h: °hue; C: Chroma.

### 3.4. Sensory Characteristics of Flakes

A preference test was conducted to determine the sensory characteristics of the flakes. The results are presented in Table 4. The highest level of color preference was found in flakes with OSP to RR ratios of 60:40, 40:60, and 20:80, while flakes containing 100% OSP had the lowest level of acceptance for color. The preference scores for taste and crispness of the flakes with various proportions of OSP and RR were comparable. Flakes containing 100% OSP and 100% RR received the highest preference score for mouthfeel, with the former having a significant edge. Therefore, mixing the OSP and RR lowers the mouthfeel acceptance of the flake products.

**Table 4.** Preference test of flakes produced from different ratios of orange sweet potato (OSP), red rice (RR).

| | Proportions of OSP and RR in Flakes | | | | | |
|---|---|---|---|---|---|---|
| | 100:0 | 80:20 | 60:40 | 40:60 | 20:80 | 0:100 |
| Color | 3.35 ± 1.42 [a] | 4.40 ± 1.22 [b] | 5.14 ± 1.12 [c] | 4.93 ± 1.21 [c] | 4.89 ± 1.30 [c] | 4.30 ± 1.12 [b] |
| Taste | 4.43 ± 1.34 [a] | 4.69 ± 1.28 [ab] | 5.08 ± 1.18 [c] | 5.09 ± 1.20 [c] | 5.05 ± 1.26 [bc] | 4.72 ± 1.29 [abc] |
| Crispness | 4.76 ± 1.16 [b] | 4.85 ± 1.04 [b] | 4.76 ± 1.40 [b] | 4.08 ± 1.17 [a] | 4.96 ± 1.07 [b] | 5.00 ± 1.04 [b] |
| Mouthfeel | 5.41 ± 0.91 [d] | 5.04 ± 1.18 [c] | 4.84 ± 1.31 [bc] | 3.91 ± 1.59 [a] | 4.66 ± 1.32 [b] | 5.05 ± 1.03 [c] |

Data are presented as mean ± standard deviation. Different superscript letters (a–d) denote significantly different values according to Duncan's test ($p < 0.05$). Comparison was made within each row.

## 4. Discussion

Phenolic compounds are the most widely found bioactive compounds in plants [33]. Some are produced in response to stress conditions as a defense mechanism of the plant. They have been extensively investigated due to their antioxidant activity and anti-degenerative disease effects [34]. In this research, methanol–HCl (1%) was used because acidic methanol can penetrate deeply into cells, disrupting the cell membrane. In addition, acidic methanol can dissolve and stabilize polar compounds such as phenolics and anthocyanins [35]. This research shows that RR and OSP are rich sources of phenolic compounds. It has previously been reported that RR has a high content of phenolic compounds [36], and that these compounds are primarily accumulated in the aleurone layer and bran of RR [37]. Thus, the rice milling process to remove the husk plays a vital role in preventing the loss of various beneficial compounds. Previously, it was suggested that ferulic acid, p-coumaric acid, and pro-catechuic acid are the most abundantly found phenolic compounds in RR [37].

Here, cooking led to a 49% decrease in the phenolic compounds of RR. Heating of RR generally destroys the structure of phenolic compounds by breaking the esterified and glycosylated bonds, thus decreasing the quantified content of phenolic compounds in cooked RR [38]. Our results agree with previous findings [39,40], which reported high levels of total phenolic compounds, mostly gallic acid, chlorogenic acid, pro-catechuic 4-hydroxybenzoic acid, and salicylic acid, in different varieties of OSP. However, phenolic contents were broken down by the heating process. Regarding the flake products, flakes containing a higher proportion of RR showed a higher content of phenolic compounds. Nevertheless, the processing lowered the phenolic compounds in flakes when compared to raw OSP and RR. Boiling and baking have previously been reported to be responsible for the loss of phenolic compounds of RR-based products [41]

A similar trend was observed in the anthocyanin content of RR. The cooking process resulted in a significant decrease in the anthocyanin content due to the unstable property of anthocyanins when exposed to high temperatures [42]. Anthocyanins were only detected in RR in this research. Anthocyanins such as cyanidin 3 glucoside, delphinidin 3 glucoside, and peonidin are reported to have health-promoting properties [43]. Thus, exposure to high temperatures for extended periods should be avoided to reduce the risk of anthocyanin breakdown. The anthocyanin content in flakes was lower than in the raw samples, and increasing the proportion of RR resulted in a higher anthocyanin content of flakes. The decrease in the anthocyanin content of flakes is possibly due to the heat treatment during flake production, typically involving two high temperature processing steps of pre-gelatinization and flaking. It has been reported that the high-temperature treatment used for food processing can lead to a reduction of the anthocyanin content [44].

Moreover, there is a growing research interest in the conversion of $\beta$-carotene and $\alpha$-carotene absorbed in the duodenum to retinol by intestinal enzymes [45]. The high rate of vitamin A deficiency in the world and the detrimental effects caused by the condition have necessitated the search for foods that can supply sufficient amounts of daily vitamin A requirement. In this research, OSP had the highest content of $\beta$-carotene. However, boiling of OSP decreased the $\beta$-carotene by approximately 41% of the $\beta$-carotene available. This phenomenon could be due to the thermal breakdown of $\beta$-carotene. Moreover, carotenoids are well known as substances that are sensitive to light and high temperatures [46]. RR contains different types of carotenoids [47]. Here, an increase in carotenoids after boiling was found, which could be related to the thermal disruption of the protein–carotenoid complex and the consequent release of carotenoids from the matrix. Similar findings have been published [48]. Similar trends were also observed in the OSP- and RR-based flake products. The $\beta$-carotene was significantly lower in the cooked sample compared to the raw sample. The OSP proportion in the flake formulation affected the carotenoid content. The higher the OSP proportion, the greater the carotenoid content of the flakes. In addition, the carotenoid contents of the flakes were significantly lower than the raw material. The simultaneous heating process from pre-gelatinization to flake pressing could further break down the carotenoids. This finding is supported by previously published work, which shows that heat treatment during food processing is responsible for the loss of carotenoids to degradation [49].

Vitamin E deficiency could lead to severe neurological problems. The main vitamin E compounds are tocopherols, with both $\alpha$- and $\gamma$-tocopherol providing most vitamin E activity. Both samples had high contents of $\alpha$- and $\gamma$-tocopherol. A significant increase (74%) in $\alpha$-tocopherol was found in OSP after boiling. This result indicates that heat treatment could be beneficial for the bioaccessibility of tocopherol. Furthermore, heat treatment can assist in the breakdown of complex foods. Thus, tocopherol can be quickly released from its binding site. On the contrary, heat treatment was reported to reduce the tocopherol content in corn [50]. The increase in tocopherol after boiling indicates that tocopherols in the sample are more stable to heat treatment than other foods. In this research, RR had a considerably high tocopherol content. Therefore, boiling could have released the tocopherols from their binding site, facilitating their extraction and detection. Moreover, the structure of the rice grains could have played a role. RR has a compact structure. Therefore, cooking could assist the extraction of tocopherols, increasing the extractable tocopherol content, although some of the tocopherols might be lost to high temperature. The tocopherol contents of the different flake products were lower than the raw and boiled OSP and RR. It could be due to the simultaneous or prolonged exposure to heat treatment. Unlike the conventional method of cooking, which increases the tocopherol content, extended heating breaks the matrix structure in flakes and significantly affects the tocopherols. Thus, prolonged heating should be avoided in the processing of healthy food products to retain their tocopherols. Heat treatment combined with mechanical treatment in specific conditions could release the tocopherols from the food matrix. However, extended exposure will lead to the breakdown of carotenoids in the sample.

This research measured the antioxidant activity of raw and cooked OSP and RR and their flake products using DPPH, FRAP, and Superoxide radical scavenging capacity assays. The result showed that heat treatment was responsible for decreasing the antioxidant activity of OSP, RR, and the flake products. A positive correlation was observed between the reduction of bioactive compounds and antioxidant capacity. Most bioactive compounds are heat sensitive, which influences their antioxidant activity. A previous study reported that an increase in temperature accelerated the initiation of oxidation, preventing antioxidant compounds from working optimally [51]. The bioactive compounds were degraded, experiencing structural changes, and wholly transformed into inactive substances. Nevertheless, due to the complex nature of antioxidant compounds in plants, their thermal stability varies. Some compounds such as pro-catechuic acid, p-coumaric acid, and ferulic acid have high thermal stability, which facilitates their extraction and antioxidant activity [52]. The heat treatment process assists in the release of such compounds without affecting their activity. Moreover, the flake products exhibited lower antioxidant activity than the raw or boiled products. Based on the percentage values of DPPH and Superoxide radical scavenging capacity and the content of Fe [II] formed, flakes containing only OSP or RR had lower antioxidant activity. Interestingly, the combination of OSP and RR increased the antioxidant activity, probably due to the synergistic effect of bioactive compounds from OSP and RR [53]. Even though the processing reduced the antioxidant activity in boiled samples and flake products, the remaining antioxidant activity was still considerably high. Therefore, boiled OSP, RR, and their flake products are a good source of bioactive compounds and antioxidants.

Regarding the physicochemical and sensory properties of the flakes, it was observed that the moisture content of flake products was mainly influenced by their composition. Starch is the dominant carbohydrate found in OSP and RR, and OSP has a lower amylose content compared to RR. According to Wang et al. [54], the amylose content of OSP is 18.71%, while Markus et al. [55] reported that RR has 23% amylose content. Amylose is a linear polymer of glucose, which forms starch. The higher the amylose content, the greater the moisture content of the dough due to a higher capability to absorb water. The absorbed water will promote dough gelatinization during heating. The water absorbed by the dough will then evaporate during the flaking process due to the network's inability to entrap water during the pre-gelatinization heat treatment. The high flaking temperature will detach water from the matrix structure of the flakes, resulting in increased evaporation and a lower moisture content of the flake products.

The dietary fiber content of the flakes ranged between 9.47 and 13.86%, comparable to values commonly found in breakfast cereals such as corn flakes, rice, quinoa, millet, and amaranth flakes [56]. The proportion of RR affected the dietary fiber of flakes. The fiber content of flakes was also influenced by the heating and pressing processes. Heat treatment caused the degradation of the fiber matrix and the glycosidic bond. The degradation affected the solubility level of the fiber, i.e., the ratio between soluble and insoluble fiber, thus resulting in the reduction of total fiber in the product.

Water absorption index (WAI) is a physical property associated with the ability of flakes to absorb water molecules within a particular time. Absorbed water molecules could be bound or detained in matrix pores of flakes. The water absorption index is crucial because it is associated with the quality of the flakes. Consumers can experience the crispy and crunchy sensation of the flakes after soaking in milk or water. In contrast, the higher WAI is interrelated with the unwanted soggy texture of flakes. The WAI is influenced by the porosity of the matrix on flakes, thickness, and hygroscopicity of flakes. In addition, the presence of fiber and protein could assist the flakes in absorbing water into their structure. The result shows that increasing the proportion of RR reduces the WAI due to a decrease in hygroscopicity. OSP is a rich source of sugar; thus, it has higher hygroscopicity compared to RR. In addition, the presence of RR affects the network construction of flakes by inhibiting the formation of the starch–protein structure, which could entrap gas. The compact structure created by RR starch inhibits water absorption into the matrix of the product. On the other hand, RR has higher fiber and protein contents, which help improve water absorption [57]. Moreover, the suitable fragmentation of the amylose and amylopectin chain in sweet potato could also affect the water absorption capacity. The heating processes such as roasting, flaking, and extrusion will induce starch fragmentation with sufficient water. The gelatinization process converts starch to a digestible material and plays a vital role in determining the structural properties of flakes and their ability to absorb moisture. The presence of RR in the flake dough disturbs the composition of starch, thus inhibiting the flakes from forming a porous structure and affecting their water absorption capacity.

Fracturability is a physical property related to deformation conditions when a specific maximum force is applied. A higher fracturability value represents the ability of food products to maintain their structure when force is applied. According to Table 2, flakes with 100% OSP have a higher fracturability value compared to others, as the homogenous matrix of starch, protein, and fiber in OSP enables the interaction between the matrix and water molecules, leading to the firm, sturdy and rigid texture of flakes. The rigid texture is related to the evolution phase of starch from amorphous conditions, which indicates complete disorganization of the crystalline structure of starch. Increasing the proportion of RR in the flakes reduces the fracturability value. The mixture of OSP and RR decreases the rigidity of the flakes due to the different structural properties of the two samples, creating flakes that are susceptible to fracture. Crispness is a complex texture attribute because it comprises a combination of sensory analysis, acoustical procedure, and instrumental analysis. The instrumental analysis revealed that flakes with OSP to RR ratios of 100:0 and 0:100 had higher crispness values than others. Similar to fracturability, the crispness value decreased with an increasing RR proportion in the flakes. The mixture of ingredients with different structural properties can affect the crispness value of flakes.

The color profile shown in Table 3 revealed that the color of flakes was affected by the pigments in the raw materials used for their production. The red color of RR is associated with anthocyanins found in its bran layer. The yellowness of OSP is linked to carotenoids, primarily β-carotene. Previous research reported that carotenoids are easily oxidized and undergo color degradation due to thermal treatment [58]. The hue results showed a flake color range between yellow and red. The color of flakes was also affected by the Maillard reaction product. The higher the Maillard reaction product, the darker the appearance of flakes.

The sensory analysis involved a preference test of the color, taste, crispness, and mouthfeel. Flakes containing 100% OSP had the lowest color preference score, associated with the orange appearance, and the lowest brightness score. Most panelists perceived the dark orange color as less fresh, less attractive, and less tasty. The color preference was increased with the addition of RR, which also helped improve the product lightness and redness values. The panelists were mostly in favor of flakes that appeared brighter and reddish. The higher brightness level is attributed to the white endosperm color of RR, while the redness is related to the anthocyanins in the bran of RR. The mouthfeel preference of flakes is associated with the water absorption index. The higher the ability of flakes to absorb water, the greater the plasticizing effect due to the presence of more hydrophilic components such as the phosphate monoester found in sweet potato starch [59]. The panelists generally preferred flakes with soft mouthfeel. The taste and crispness preferences for OSP- and RR-based flakes were in the range of "indifferent" and "slightly likes". Increasing the RR proportion in flakes decreases the bitterness intensity and increases the savory taste. The perception of savory taste is generally influenced by the moisture content and the flavor of RR. Niu et al. [60] suggests that sweet potato with a low dry matter content has a bitter taste, and increasing the dry matter reduces the bitterness. On the other hand, a decrease in invertase activity may support the bitter aftertaste of the sweet potato.

This research successfully monitored the changes of the bioactive compound and antioxidant activity of OSP and RR in their native form and in the flake products. It can be observed that individual compounds acted differently to processing methods. Therefore, it can be suggested that research on the development of functional foods should address the products ready to be consumed instead of solely focusing on the raw materials due to the changes that take place during the transformation. This approach should be implemented for other potential materials rich in bioactive compounds. Moreover, further consideration of the bioaccessibility and bioavailability that are affected by digestion and absorption in the human metabolism system should also be considered [61].

## 5. Conclusions

OSP and RR are rich sources of bioactive compounds, especially β-carotene, for OSP, and phenolic compounds and anthocyanins, for RR. The boiling process significantly decreased most of the bioactive compounds, except tocopherols and α-carotene. The level of bioactive compounds in the flake products was dependent on the proportion of OSP and RR. Heat treatment resulted in a decrease in antioxidant activity, even though the remaining activity was still considerably high. The mixture of OSP and RR can produce flakes with low moisture and high fiber contents. The optimum flake water absorption index, fracturability, and crispness were obtained by combining 40% OSP and 60% RR. Moreover, the ratio of OSP and RR influenced the color and sensory preferences of the panelists.

**Author Contributions:** Conceptualization, I.R.A.P.J.; methodology, I.R.A.P.J., S.R., T.I.P.S. and L.M.Y.D.D.; formal analysis, I.R.A.P.J. and C.W.; investigation, S.R. and I.R.A.P.J.; resources, L.M.Y.D.D. and T.I.P.S.; data curation, I.R.A.P.J. and L.M.Y.D.D.; writing—original draft preparation, I.R.A.P.J. and L.M.Y.D.D.; writing—review and editing, C.W.; visualization, I.R.A.P.J.; supervision, C.W.; project administration, I.R.A.P.J. and S.R.; funding acquisition, I.R.A.P.J., S.R. and T.I.P.S. All authors have read and agreed to the published version of the manuscript.

**Funding:** This research was funded by The Ministry of Research, Technology, and Higher Education, Republic of Indonesia, grant number 130U/WM01.5/N/2020. The APC was also funded by The Ministry of Research, Technology, and Higher Education, Republic of Indonesia as part of the research grant and Widya Mandala Surabaya Catholic University (6477/WM01/C/2022).

**Institutional Review Board Statement:** Not applicable.

**Informed Consent Statement:** Not applicable.

**Data Availability Statement:** Data is contained within the article.

**Conflicts of Interest:** The authors declare no conflict of interest.

## References

1. Shahidi, F.; Yeo, J. Bioactivities of Phenolics by Focusing on Suppression of Chronic Diseases: A Review. *Int. J. Mol. Sci.* **2018**, *19*, 1573. [CrossRef] [PubMed]
2. Shao, Y.; Xu, F.; Sun, X.; Bao, J.; Beta, T. Identification and Quantification of Phenolic Acids and Anthocyanins as Antioxidants in Bran, Embryo and Endosperm of White, Red and Black Rice Kernels (*Oryza Sativa* L.). *J. Cereal Sci.* **2014**, *59*, 211–218. [CrossRef]
3. Bhat, F.M.; Sommano, S.R.; Riar, C.S.; Seesuriyachan, P.; Chaiyaso, T.; Prom-u-Thai, C. Status of Bioactive Compounds from Bran of Pigmented Traditional Rice Varieties and Their Scope in Production of Medicinal Food with Nutraceutical Importance. *Agronomy* **2020**, *10*, 1817. [CrossRef]
4. Rathna Priya, T.S.; Eliazer Nelson, A.R.L.; Ravichandran, K.; Antony, U. Nutritional and Functional Properties of Coloured Rice Varieties of South India: A Review. *J. Ethn. Food* **2019**, *6*, 11. [CrossRef]
5. Upanan, S.; Yodkeeree, S.; Thippraphan, P.; Punfa, W.; Wongpoomchai, R.; Limtrakul, P. The Proanthocyanidin-Rich Fraction Obtained from Red Rice Germ and Bran Extract Induces HepG2 Hepatocellular Carcinoma Cell Apoptosis. *Molecules* **2019**, *24*, 813. [CrossRef]
6. Kim, T.J.; Kim, S.Y.; Park, Y.J.; Lim, S.-H.; Ha, S.-H.; Park, S.U.; Lee, B.; Kim, J.K. Metabolite Profiling Reveals Distinct Modulation of Complex Metabolic Networks in Non-Pigmented, Black, and Red Rice (*Oryza Sativa* L.) Cultivars. *Metabolites* **2021**, *11*, 367. [CrossRef]
7. Samyor, D.; Deka, S.C.; Das, A.B. Effect of Extrusion Conditions on the Physicochemical and Phytochemical Properties of Red Rice and Passion Fruit Powder Based Extrudates. *J. Food Sci. Technol.* **2018**, *55*, 5003–5013. [CrossRef]

8.   Müller, C.P.; Hoffmann, J.F.; Ferreira, C.D.; Diehl, G.W.; Rossi, R.C.; Ziegler, V. Effect of Germination on Nutritional and Bioactive Properties of Red Rice Grains and Its Application in Cupcake Production. *Int. J. Gastron. Food Sci.* **2021**, *25*, 100379. [CrossRef]

9.   Mbanjo, E.G.N.; Kretzschmar, T.; Jones, H.; Ereful, N.; Blanchard, C.; Boyd, L.A.; Sreenivasulu, N. The Genetic Basis and Nutritional Benefits of Pigmented Rice Grain. *Front. Genet.* **2020**, *11*, 229. [CrossRef]

10.  Cappa, C.; Laureati, M.; Casiraghi, M.C.; Erba, D.; Vezzani, M.; Lucisano, M.; Alamprese, C. Effects of Red Rice or Buckwheat Addition on Nutritional, Technological, and Sensory Quality of Potato-Based Pasta. *Foods* **2021**, *10*, 91. [CrossRef]

11.  Kraithong, S.; Lee, S.; Rawdkuen, S. Effect of Red Jasmine Rice Replacement on Rice Flour Properties and Noodle Qualities. *Food Sci. Biotechnol.* **2019**, *28*, 25–34. [CrossRef]

12.  Meza, S.L.R.; Sinnecker, P.; Schmiele, M.; Massaretto, I.L.; Chang, Y.K.; Marquez, U.M.L. Production of Innovative Gluten-Free Breakfast Cereals Based on Red and Black Rice by Extrusion Processing Technology. *J. Food Sci. Technol.* **2019**, *56*, 4855–4866. [CrossRef] [PubMed]

13.  Deeseenthum, S.; Luang-In, V.; Chunchom, S. Characteristics of Thai Pigmented Rice Milk Kefirs with Potential as Antioxidant and Anti-Inflammatory Foods. *Pharmacogn. J.* **2017**, *10*, 154–161. [CrossRef]

14.  Sulistyaningtyas, A.R.; Lunggani, A.T.; Kusdiyantini, E. Kefir Produced from Red Rice Milk by Lactobacillus Bulgaricus and Candida Kefir Starter. *IOP Conf. Ser. Earth Environ. Sci.* **2019**, *292*, 012038. [CrossRef]

15.  Kourouma, V.; Mu, T.-H.; Zhang, M.; Sun, H.-N. Effects of Cooking Process on Carotenoids and Antioxidant Activity of Orange-Fleshed Sweet Potato. *LWT* **2019**, *104*, 134–141. [CrossRef]

16.  Sebben, J.A.; da Silveira Espindola, J.; Ranzan, L.; Fernandes de Moura, N.; Trierweiler, L.F.; Trierweiler, J.O. Development of a Quantitative Approach Using Raman Spectroscopy for Carotenoids Determination in Processed Sweet Potato. *Food Chem.* **2018**, *245*, 1224–1231. [CrossRef]

17.  Alam, M.K.; Sams, S.; Rana, Z.H.; Akhtaruzzaman, M.; Islam, S.N. Minerals, Vitamin C, and Effect of Thermal Processing on Carotenoids Composition in Nine Varieties Orange-Fleshed Sweet Potato (Ipomoea Batatas L.). *J. Food Compos. Anal.* **2020**, *92*, 103582. [CrossRef]

18.  Low, J.W.; Mwanga, R.O.M.; Andrade, M.; Carey, E.; Ball, A.-M. Tackling Vitamin A Deficiency with Biofortified Sweetpotato in Sub-Saharan Africa. *Glob. Food Secur.* **2017**, *14*, 23–30. [CrossRef]

19.  Laryea, D.; Wireko-Manu, F.D.; Oduro, I. Formulation and Characterization of Sweetpotato-Based Complementary Food. *Cogent Food Agric.* **2018**, *4*, 1517426. [CrossRef]

20.  Neela, S.; Fanta, S.W. Review on Nutritional Composition of Orange-fleshed Sweet Potato and Its Role in Management of Vitamin A Deficiency. *Food Sci. Nutr.* **2019**, *7*, 1920–1945. [CrossRef]

21.  Olagunju, A.I.; Omoba, O.S.; Awolu, O.O.; Rotowa, K.O.; Oloniyo, R.O.; Ogunowo, O.C. Physiochemical, Antioxidant Properties and Carotenoid Retention/Loss of Culinary Processed Orange Fleshed Sweet Potato. *J. Culin. Sci. Technol.* **2020**, *19*, 1–20. [CrossRef]

22.  Pan, Z.; Sun, Y.; Zhang, F.; Guo, X.; Liao, Z. Effect of Thermal Processing on Carotenoids and Folate Changes in Six Varieties of Sweet Potato (*Ipomoes Batata* L.). *Foods* **2019**, *8*, 215. [CrossRef]

23.  Klepacka, J.; Najda, A. Effect of Commercial Processing on Polyphenols and Antioxidant Activity of Buckwheat Seeds. *Int. J. Food Sci. Technol.* **2021**, *56*, 661–670. [CrossRef]

24.  Santos, D.I.; Saraiva, J.M.A.; Vicente, A.A.; Moldão-Martins, M. Methods for Determining Bioavailability and Bioaccessibility of Bioactive Compounds and Nutrients. In *Innovative Thermal and Non-Thermal Processing, Bioaccessibility and Bioavailability of Nutrients and Bioactive Compounds*; Elsevier: Amsterdam, The Netherlands, 2019; pp. 23–54. ISBN 978-0-12-814174-8.

25.  Radix, A.P.; Jati, I.; Nohr, D.; Biesalski, H.K. Nutrients and Antioxidant Properties of Indonesian Underutilized Colored Rice. *Nutr. Food Sci.* **2014**, *44*, 193–203. [CrossRef]

26.  Astadi, I.R.; Astuti, M.; Santoso, U.; Nugraheni, P.S. In Vitro Antioxidant Activity of Anthocyanins of Black Soybean Seed Coat in Human Low Density Lipoprotein (LDL). *Food Chem.* **2009**, *112*, 659–663. [CrossRef]

27.  Giusti, M.M.; Wrolstad, R.E. Characterization and Measurement of Anthocyanins by UV-Visible Spectroscopy. *Curr. Protoc. Food Anal. Chem.* **2001**, *00*, F1–F2. [CrossRef]

28.  Gautam, B.; Vadivel, V.; Stuetz, W.; Biesalski, H.K. Bioactive Compounds Extracted from Indian Wild Legume Seeds: Antioxidant and Type II Diabetes-Related Enzyme Inhibition Properties. *Int. J. Food Sci. Nutr.* **2012**, *63*, 242–245. [CrossRef]

29.  Vadivel, V.; Biesalski, H.K. Contribution of Phenolic Compounds to the Antioxidant Potential and Type II Diabetes Related Enzyme Inhibition Properties of Pongamia Pinnata L. Pierre Seeds. *Process Biochem.* **2011**, *46*, 1973–1980. [CrossRef]

30.  Obi, O.F.; Ezeoha, S.L.; Egwu, C.O. Evaluation of Air Oven Moisture Content Determination Procedures for Pearl Millet (*Pennisetum Glaucum* L.). *Int. J. Food Prop.* **2016**, *19*, 454–466. [CrossRef]

31.  Hu, X.-Z.; Zheng, J.-M.; Li, X.; Xu, C.; Zhao, Q. Chemical Composition and Sensory Characteristics of Oat Flakes: A Comparative Study of Naked Oat Flakes from China and Hulled Oat Flakes from Western Countries. *J. Cereal Sci.* **2014**, *60*, 297–301. [CrossRef]

32.  Salawi, A.; Nazzal, S. The Rheological and Textural Characterization of Soluplus®/Vitamin E Composites. *Int. J. Pharm.* **2018**, *546*, 255–262. [CrossRef] [PubMed]

33.  Karatas, N.; Sengul, M. Some Important Physicochemical and Bioactive Characteristics of the Main Apricot Cultivars from Turkey. *Turk. J. Agric. For.* **2020**, *44*, 651–661. [CrossRef]

34. Luna-Guevara, M.L.; Luna-Guevara, J.J.; Hernández-Carranza, P.; Ruíz-Espinosa, H.; Ochoa-Velasco, C.E. Phenolic Compounds: A Good Choice Against Chronic Degenerative Diseases. In *Studies in Natural Products Chemistry*; Elsevier: Amsterdam, The Netherlands, 2018; Volume 59, pp. 79–108, ISBN 978-0-444-64179-3.
35. Horvat, D.; Šimić, G.; Drezner, G.; Lalić, A.; Ledenčan, T.; Tucak, M.; Plavšić, H.; Andrić, L.; Zdunić, Z. Phenolic Acid Profiles and Antioxidant Activity of Major Cereal Crops. *Antioxidants* **2020**, *9*, 527. [CrossRef]
36. Ghasemzadeh, A.; Karbalaii, M.T.; Jaafar, H.Z.E.; Rahmat, A. Phytochemical Constituents, Antioxidant Activity, and Antiproliferative Properties of Black, Red, and Brown Rice Bran. *Chem. Cent. J.* **2018**, *12*, 17. [CrossRef] [PubMed]
37. Pang, Y.; Ahmed, S.; Xu, Y.; Beta, T.; Zhu, Z.; Shao, Y.; Bao, J. Bound Phenolic Compounds and Antioxidant Properties of Whole Grain and Bran of White, Red and Black Rice. *Food Chem.* **2018**, *240*, 212–221. [CrossRef]
38. Fracassetti, D.; Pozzoli, C.; Vitalini, S.; Tirelli, A.; Iriti, M. Impact of Cooking on Bioactive Compounds and Antioxidant Activity of Pigmented Rice Cultivars. *Foods* **2020**, *9*, 967. [CrossRef]
39. Azeem, M.; Mu, T.-H.; Zhang, M. Influence of Particle Size Distribution of Orange-Fleshed Sweet Potato Flour on Dough Rheology and Simulated Gastrointestinal Digestion of Sweet Potato-Wheat Bread. *LWT* **2020**, *131*, 109690. [CrossRef]
40. Ruttarattanamongkol, K.; Chittrakorn, S.; Weerawatanakorn, M.; Dangpium, N. Effect of Drying Conditions on Properties, Pigments and Antioxidant Activity Retentions of Pretreated Orange and Purple-Fleshed Sweet Potato Flours. *J. Food Sci. Technol.* **2016**, *53*, 1811–1822. [CrossRef]
41. Ziegler, V.; Ferreira, C.D.; Hoffmann, J.F.; Chaves, F.C.; Vanier, N.L.; de Oliveira, M.; Elias, M.C. Cooking Quality Properties and Free and Bound Phenolics Content of Brown, Black, and Red Rice Grains Stored at Different Temperatures for Six Months. *Food Chem.* **2018**, *242*, 427–434. [CrossRef]
42. Sasongko, S.B.; Djaeni, M.; Utari, F.D. Kinetic of Anthocyanin Degradation in Roselle Extract Dried with Foaming Agent at Different Temperatures. *Bull. Chem. React. Eng. Catal.* **2019**, *14*, 320. [CrossRef]
43. Zhu, F. Anthocyanins in Cereals: Composition and Health Effects. *Food Res. Int.* **2018**, *109*, 232–249. [CrossRef]
44. Meléndez-Martínez, A.J. An Overview of Carotenoids, Apocarotenoids, and Vitamin A in Agro-Food, Nutrition, Health, and Disease. *Mol. Nutr. Food Res.* **2019**, *63*, 1801045. [CrossRef] [PubMed]
45. Trono, D. Carotenoids in Cereal Food Crops: Composition and Retention throughout Grain Storage and Food Processing. *Plants* **2019**, *8*, 551. [CrossRef] [PubMed]
46. Oduro-Obeng, H.; Fu, B.X.; Beta, T. Influence of Cooking Duration on Carotenoids, Physical Properties and in Vitro Antioxidant Capacity of Pasta Prepared from Three Canadian Durum Wheat Cultivars. *Food Chem.* **2021**, *363*, 130016. [CrossRef] [PubMed]
47. Melini, V.; Panfili, G.; Fratianni, A.; Acquistucci, R. Bioactive Compounds in Rice on Italian Market: Pigmented Varieties as a Source of Carotenoids, Total Phenolic Compounds and Anthocyanins, before and after Cooking. *Food Chem.* **2019**, *277*, 119–127. [CrossRef]
48. Nartea, A.; Fanesi, B.; Falcone, P.M.; Pacetti, D.; Frega, N.G.; Lucci, P. Impact of Mild Oven Cooking Treatments on Carotenoids and Tocopherols of Cheddar and Depurple Cauliflower (Brassica Oleracea L. Var. Botrytis). *Antioxidants* **2021**, *10*, 196. [CrossRef]
49. Tang, Y.; Cai, W.; Xu, B. Profiles of Phenolics, Carotenoids and Antioxidative Capacities of Thermal Processed White, Yellow, Orange and Purple Sweet Potatoes Grown in Guilin, China. *Food Sci. Hum. Wellness* **2015**, *4*, 123–132. [CrossRef]
50. Da Silva Timm, N.; Ramos, A.H.; Ferreira, C.D.; de Oliveira Rios, A.; Zambiazi, R.C.; de Oliveira, M. Influence of Germ Storage from Different Corn Genotypes on Technological Properties and Fatty Acid, Tocopherol, and Carotenoid Profiles of Oil. *Eur. Food Res. Technol.* **2021**, *247*, 1449–1460. [CrossRef]
51. Bagchi, T.B.; Chattopadhyay, K.; Sivashankari, M.; Roy, S.; Kumar, A.; Biswas, T.; Pal, S. Effect of Different Processing Technologies on Phenolic Acids, Flavonoids and Other Antioxidants Content in Pigmented Rice. *J. Cereal Sci.* **2021**, *100*, 103263. [CrossRef]
52. Călinoiu, L.; Vodnar, D. Thermal Processing for the Release of Phenolic Compounds from Wheat and Oat Bran. *Biomolecules* **2019**, *10*, 21. [CrossRef]
53. Sitarek, P.; Merecz-Sadowska, A.; Kowalczyk, T.; Wieczfinska, J.; Zajdel, R.; Śliwiński, T. Potential Synergistic Action of Bioactive Compounds from Plant Extracts against Skin Infecting Microorganisms. *Int. J. Mol. Sci.* **2020**, *21*, 5105. [CrossRef] [PubMed]
54. Wang, H.; Yang, Q.; Ferdinand, U.; Gong, X.; Qu, Y.; Gao, W.; Ivanistau, A.; Feng, B.; Liu, M. Isolation and Characterization of Starch from Light Yellow, Orange, and Purple Sweet Potatoes. *Int. J. Biol. Macromol.* **2020**, *160*, 660–668. [CrossRef] [PubMed]
55. Markus, J.E.R.; Ndiwa, A.S.S.; Oematan, S.S.; Mau, Y.S. Variations of Grain Physical Properties, Amylose and Anthocyanin of Upland Red Rice Cultivars from East Nusa Tenggara, Indonesia. *Biodiversitas* **2021**, *22*, 1345–1354. [CrossRef]
56. Gates, F. *Role of Heat Treatment in the Processing and Quality of Oat Flakes*; University of Helsinki: Helsinki, Finland, 2007; ISBN 978-952-10-3907-2.
57. Zhou, Y.; Dhital, S.; Zhao, C.; Ye, F.; Chen, J.; Zhao, G. Dietary Fiber-Gluten Protein Interaction in Wheat Flour Dough: Analysis, Consequences and Proposed Mechanisms. *Food Hydrocoll.* **2021**, *111*, 106203. [CrossRef]
58. Trancoso-Reyes, N.; Ochoa-Martínez, L.A.; Bello-Pérez, L.A.; Morales-Castro, J.; Estévez-Santiago, R.; Olmedilla-Alonso, B. Effect of Pre-Treatment on Physicochemical and Structural Properties, and the Bioaccessibility of β-Carotene in Sweet Potato Flour. *Food Chem.* **2016**, *200*, 199–205. [CrossRef] [PubMed]
59. Akintayo, O.A.; Obadu, J.M.; Karim, O.R.; Balogun, M.A.; Kolawole, F.L.; Oyeyinka, S.A. Effect of Replacement of Cassava Starch with Sweet Potato Starch on the Functional, Pasting and Sensory Properties of Tapioca Grits. *LWT* **2019**, *111*, 513–519. [CrossRef]

60. Niu, S.; Li, X.-Q.; Tang, R.; Zhang, G.; Li, X.; Cui, B.; Mikitzel, L.; Haroon, M. Starch Granule Sizes and Degradation in Sweet Potatoes during Storage. *Postharvest Biol. Technol.* **2019**, *150*, 137–147. [CrossRef]
61. Granato, D.; Shahidi, F.; Wrolstad, R.; Kilmartin, P.; Melton, L.D.; Hidalgo, F.J.; Miyashita, K.; van Camp, J.; Alasalvar, C.; Ismail, A.B.; et al. Antioxidant Activity, Total Phenolics and Flavonoids Contents: Should We Ban in Vitro Screening Methods? *Food Chem.* **2018**, *264*, 471–475. [CrossRef]

*Review*

# Undervalued Spiny Monkey Orange (*Strychnos spinosa* Lam.): An Indigenous Fruit for Sustainable Food-Nutrition and Economic Prosperity

Abiodun Olusola Omotayo [1],* and Adeyemi Oladapo Aremu [1,2],*

1 Food Security and Safety Niche Area, Faculty of Natural and Agricultural Sciences, North-West University, Private Bag X2046, Mmabatho 2745, South Africa
2 Indigenous Knowledge Systems Centre, Faculty of Natural and Agricultural Sciences, North-West University, Private Bag X2046, Mmabatho 2745, South Africa
* Correspondence: 25301284@nwu.ac.za (A.O.O.); Oladapo.Aremu@nwu.ac.za (A.O.A.); Tel.: +27-18-389-2573 (A.O.O.)

**Abstract:** *Strychnos spinosa* Lam. is among the top nutrient-dense indigenous fruit species that are predominant in Southern Africa. It is a highly ranked indigenous fruit based on the nutrition and sensorial properties, which make it an important food source for the marginalized rural people. On the basis of the high vitamin C, iron, and zinc content, it has the capacity to improve the food-nutrition and the socioeconomic status of individuals, especially those in the rural areas of the developing nations. The nutritional composition of *Strychnos spinosa* compare favorably with many of the popular fruits, such as strawberries and orange. Additionally, *Strychnos spinosa* has antioxidant activity similar to well-known antioxidant fruits, which keeps it in the class of the popular fruits, giving it added nutrition–health-promoting benefits. In order to improve the availability of *Strychnos spinosa*, more research on the domestication, processing, preservation, value chain, and economic potential need to be further explored. Therefore, we recommend more concerted efforts from relevant stakeholders with interest in *Strychnos spinosa* fruit production as a possible sustainable solution to food shortage, food-nutrition insecurity, malnutrition, and austerity, mainly in the rural communities of the developing countries.

**Keywords:** fruit tree; food policies; food security; Loganiaceae; nutrients; market economies; novel products

**Citation:** Omotayo, A.O.; Aremu, A.O. Undervalued Spiny Monkey Orange (*Strychnos spinosa* Lam.): An Indigenous Fruit for Sustainable Food-Nutrition and Economic Prosperity. *Plants* **2021**, *10*, 2785. https://doi.org/10.3390/plants10122785

Academic Editor: Ivo Vaz de Oliveira

Received: 19 October 2021
Accepted: 9 December 2021
Published: 16 December 2021

**Publisher's Note:** MDPI stays neutral with regard to jurisdictional claims in published maps and institutional affiliations.



## 1. Introduction

With the declined rate of global undernourishment (15% within 2000–2004 and 8.9% in the year 2019), about 690 million individuals remain undernourished globally. Meanwhile, the stunting rate further fell from 33% of children under age five in 2000 to 21.3% in 2019 [1–3]. In order to achieve the goal of ending undernutrition by the year 2030, there is need to encourage the consumption of a balanced diet, especially in the rural communities of the developing nations [4–7]. Interestingly, existing literature have ascertained that indigenous fruits are used to cover food lack and shortages, thereby, these remain a key option for dealing with micronutrient shortages during vulnerable times [8,9]. Indigenous fruits have been utilized in several ways since time immemorial for food needs of the local societies [3].

Generally, the potential of many indigenous fruits is underexplored, especially in the area of their basic botany, horticulture, food science, and economic value [10–13]. Indigenous fruits have the potential to provide the necessary phytonutrients required in the diet for food-nutrition security and the income of rural communities where the cultivation of the popular fruit species is not common [14]. In the warmer temperate regions of the globe, an indigenous fruit tree that stands out with a rich source of phytonutrients is the

*Strychnos spinosa* [15]. It is one of the most important edible indigenous fruit trees in the wild. The fruit-bearing species of *Strychnos* belong to the family Loganiaceae. The tree has the capacity to stay edible in tropical heat, which is an important characteristic for food and nutrition security, as this will enhance availability and productivity [16–18].

In traditional medicine, *Strychnos spinosa* is often used in the treatment of venereal diseases, stomach-related aches, and snake bite attack [19]. *Strychnos spinosa* is known as a native or introduced species in many African nations. The plant has been reported across different African regions, including Southern Africa, East Africa, and West Africa [20]. In South Africa, *Strychnos spinosa* grows well in four provinces (Eastern Cape, Limpopo, KwaZulu-Natal, and Mpumalanga).

Furthermore, the conservation status of *Strychnos spinosa* is categorized as "least concern", as its distribution and abundance possess a low risk of extinction [21,22]. However, the plant has a recent record of declining occurrence in Benin and Burkina Faso (West Africa), which was attributed to factors such as agricultural activities, urbanization, and animal breeding, rather than climate change and its impact [23]. Although the distribution and availability of the *Strychnos spinosa* is uneven in Africa, its food-nutritional and economic potentials suggest the need for a more conscious and holistic conservation approach.

*Strychnos spinosa* has several local uses, and it is known to be a rich source of nutrition and phytochemicals, thereby suggesting its potential health benefits [21,24,25]. Given the increasing importance of *Strychnos spinosa* in food-nutritional sovereignty, as well as its ecological advantage [26–28], this review provides an appraisal on the potential for sustainable food–nutrition and economic prosperity of *Strychnos spinosa*. It is anticipated that consolidated information on *Strychnos spinosa* is important in an attempt to unfold its nutritional and economic potential.

## 2. Method for Literature Search

The approach described by Omotayo et al. [29] was employed in literature selection. Different online sources, theses, dissertations, and research reports were explored. We searched online sources such as Web of Science (WOS), Google Scholar, PubMeb, and Scopus, using various terms and phrases. Examples of these include "*Strychnos spinosa*", "Monkey orange", "nutritional value composition *Strychnos spinosa*", "ethno-medicinal importance of the *Strychnos spinosa*", "uses of *Strychnos spinosa*", and "description of *Strychnos spinosa*". For this review, the focus of the search was on Africa, southern Africa, and South Africa from the year 1962 to December 2021.

For the search, studies that fit the inclusion criteria were derived in order to explore the content. The five areas explored and categorized were (i) distribution and description of *Strychnos spinosa*, (ii) uses of *Strychnos spinosa*, nutritional and phytochemical content (iii) economic potential (iv) postharvest handling, preservation, storage and processing, and (v) domestication of *Strychnos spinosa*, cultivation problems, and future research direction (Table 1). In this review, a sum of 151 peer-reviewed papers were retrieved that focused on *Strychnos spinosa*. Finally, an estimated 47.68% (72) of the literature was relevant, utilized, and included in the review article (Figure 1).

**Table 1.** Selection criteria applied for the selection of literature in this review.

| Exclusion | Description |
| --- | --- |
| Underutilized African fruit plants | Existing studies on different edible and non-edible fruit plants |
| Underutilized southern African fruit plants | Literature on different edible indigenuos fruits of southern Africa |
| History and horticulture | Resarch publications on origin, taxonomy, morphology, uses, domestication, and cultivation of indigenous fruits |
| Chemical composition | Papers on the chemical composition and use of indigenous fruits |
| Non-edible uses | Literature describing uses of indigenous fruits |
| Inclusion | Explanation |
| Main subject is food nutrition and economic potential of *Strychnos spinosa* fruit tree | Nutrition literature, uses, chemicals, and prospects of *Strychnos spinosa* |
| Description, distribution, and ecology of *Strychnos spinosa* | Articles on distribution, taxonomy, morphology, and distribution of *Strychnos spinosa* |
| Diverse uses of *Strychnos spinosa* | Articles documenting the uses of *Strychnos spinosa* |
| Nutritional and phytochemical content of *Strychnos spinosa* | Nutritional, phytochemicals contents of *Strychnos spinosa* |
| Postharvest handling, preservation, storage, and processing of *Strychnos spinosa* | Articles on postharvest, preservation, and processing of *Strychnos spinosa* |
| Challenges, domestication of *Strychnos spinosa*, cultivation problems, and way forward | Domestication of *Strychnos spinosa*. Articles on food value chain, trade, economic prospects of plants, markets, supply chains, policy, and interventions. |

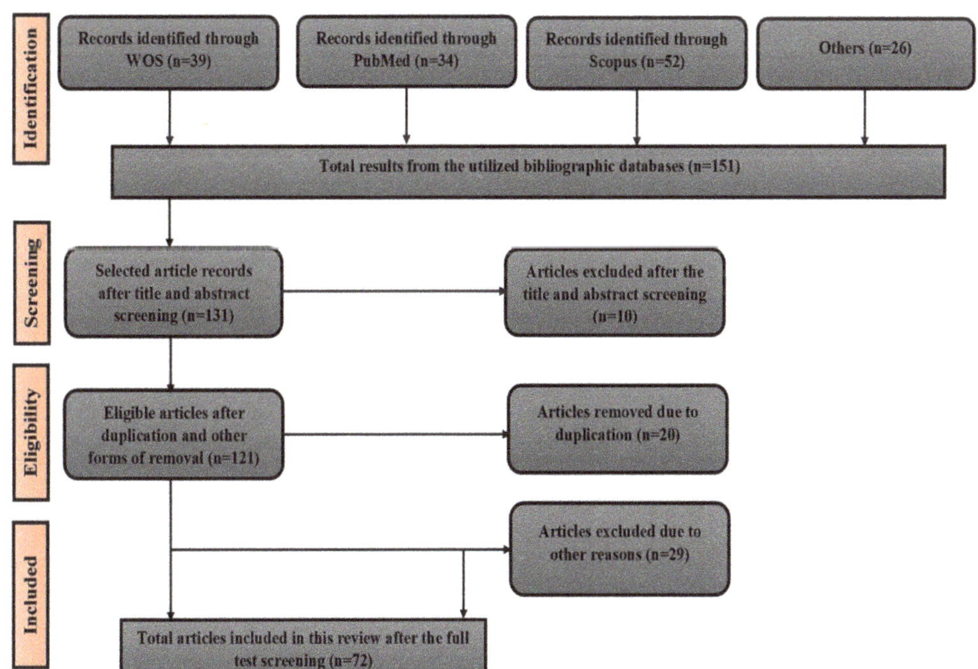


**Figure 1.** Preferred reporting items for systematic reviews and meta-analyses (PRISMA) for the exclusion and inclusion of articles.

### 3. Botanical Description and Taxonomy of *Strychnos spinosa*

About 75 species of *Strychnos* exist in Africa, with 20 species (e.g., *Strychnos innocua*, *Strychnos cocculoides*, *Strychnos pungens*, and *Strychnos spinosa*) producing consumable fruits in drought-prone and semi-arid areas [18,30,31]. *Strychnos spinosa* is a small tree of 1–7 m height, having straight and curved axillary spines, as well as a corky back [32]. The leaves are simple and oval (Figure 2a,b). The fruit is edible, round-shaped, 6–15 cm in diameter, and resembles a typical orange [31,33]. The unripe fruits (Figure 2b) are green, with wood peel of 34 mm that becomes yellow (Figure 2c) when ripe [31].

**Figure 2.** Morphology of *Strychnos spinosa*. (**a**) tree at fruiting stage; (**b**) mature green fruit; (**c**) ripe fruit.

*Strychnos spinosa* fruit has a juicy, sweet-sour pulp, which is pale brown, with about a 3 cm flat seed, slightly similar to apricots [34]. *Strychnos spinosa* grows in well-drained soils [33,35]. Fruit weighs between 145 and 383 g, while about 300–700 fruits (40–100 kg) can be produced per tree stand. *Strychnos spinosa* is a seasonal fruit tree that is harvested between August and December [31]. However, the domestication of *Strychnos spinosa* remains in experimental stages, which is still a problem associated with its commercial prospect. Presently, *Strychnos spinosa* can be propagated via seeds, grafting, or budding, with the production of fruit starting 3–5 years after planting [27].

### 4. Nutritional and Phytochemical Content of *Strychnos spinosa*

*4.1. Nutritional Composition of the Strychnos spinosa*

*Strychnos spinosa* fruit contain energy, fibers, crude protein, and minerals (Table 2) [18]. Compared to other fruits, the vitamin C content for *Strychnos spinosa* is similar to that of oranges (*Citrus sinensis*) (50 mg/100 g) and strawberries (*Fragaria ananassa*) (59 mg/100 g) [18,31]. Therefore, the consumption of *Strychnos spinosa* provides a source of ascorbate and may alleviate nutrition insecurity for local communities. Most importantly, its fruit pulp (Figure 2c) can be sun-dried as a food preserve, thereby extending shelf-life and availability.

**Table 2.** Proximate, vitamin C, and mineral composition of *Strychnos spinosa* fruit.

| Component | Content Based on Amarteifio and Mosase [36] |
|---|---|
| Proximate and vitamin C composition | |
| Dry matter | 19.7 (%) |
| Ash | 4.6 (%) |
| Crude protein | 3.3 (%) |
| Fat | na |
| Fibre | na |
| Acid detergent lignin | 4.4 (%) |
| Acid detergent fibre | 6.1 (%) |
| Neutral detergent fibre | 6.2 (%) |
| Total carbohydrate | na |
| Energy value (kJ/100 g) | na |
| Vitamin C | 88 (mg/100 g) |
| Total soluble sugar (%) | na |
| Total sugar | na |
| Total acidity | na |
| Mineral composition (mg/100 g FW) | |
| Phosphorus | 66 |
| Calcium | 56 |
| Magnesium | 49 |
| Iron | 0.11 |
| Potassium | 1370 |
| Sodium | 21.7 |
| Zinc | 0.22 |
| Copper | na |
| Manganese | na |

Note: na = not available, FW = fresh weight.

*Strychnos spinosa* fruit is a good dietary source of carbohydrates and proteins. Furthermore, it contains important minerals, namely iron, zinc, copper, and manganese [37], thereby suggesting that the consumption of *Strychnos spinosa* may serve as a source to meet the body requirement of zinc, iron, copper, and manganese. The deficiency of microminerals in the human body impairs growth and increases the susceptibility of such individuals to infections and risk of mortality, especially in children [38]. Although the presence of these aforementioned minerals in *Strychnos spinosa* fruit has been indicated, a wide variability in concentrations for some of them as reported by Lockett, et al. [39].

*4.2. Phytochemicals in Strychnos spinosa*

Phytochemicals are biological active compounds, such as the flavonoids and phenolic acids, with health-promoting values, such as anti-ageing and inflammation [18,21,40,41], which were mainly attributed to their ability to scavenge free radicals [18,42,43]. The rich phytochemicals that are abound in different parts of *Strychnos spinosa* remain key to explaining their food-nutritional benefits and future potential [44–47]. Diverse phytochemicals were confirmed in the leaves, branches, seeds, and fruit pericarp of *Strychnos spinosa* (Table 3). In addition, significant amount of phenolics and flavonoids were detected in the root-bark [37,48].

**Table 3.** Overview of phytochemicals in *Strychnos spinosa*.

| Plant Part | Examples of Phytochemical |
|---|---|
| Leaves | Glycosides, tannins, saponins, anthraquinones, steroids, alkaloids, and terpenoids [24,47,49] |
| Branches | Tannins, flavonoids, terpenoids, saponin, steroids, glycosides, and phenols [50,51] |
| Stem bark | Tannins, saponins, anthraquinones, steroids, alkaloids, glycosides, and terpenoids [24] |
| Seed | Alkaloids, tannins, phenols, phlobatannins, and steroids [52] |
| Fruit pericarp | Alkaloids, terpenes, sterols, fatty acids, flavonoids, and saponin [53] |
| Root-bark | Alkaloids, glycosides, steroids and terpenoids, tannins, anthraquinones, phlobatannins, and saponins [37,50] |

*4.3. Physicochemical Properties of Strychnos spinosa*

*Strychnos spinosa* fruit shows a delicate complex of aroma volatiles that are identified as a mixture of apricot, clove, pineapple, and citrus [26,33]. The degree of *Strychnos spinosa* ripeness influences the taste and sugar profile that varies based on the environmental-related factors [18]. Based on existing studies (Table 4), a wide variation have been confirmed in *Strychnos spinosa* [18,31]. The presence of organic acids in *Strychnos spinosa* is explained by the acidic content that blends with sugars, thereby making the plant to exert a blended acid-sweet taste [18]. The partial solubilization of the pectin and cellulose by the plants' enzymes, polygalacturonase [54], pectinmethylesterase, and lyase, during ripening affects the texture and juiciness of the fruit [18,31]. The sensory studies reveal that potential exists for product development and commercialization of the plant.

**Table 4.** Sensory properties in the *Strychnos spinosa* fruit.

| Properties | Description |
|---|---|
| Taste | Tarty/fermented acid-sweet [18,24] |
| Aroma volatiles | Major compound (>75%): trans-isoeugenol—4.762 mg/g FW [18,24]<br>Other compounds: eugenol—307 µg/g FW; chavicol—172 µg/g FW;<br>ρ-trans-anol—647.5 µg/g FW; 123.5 µg/g FW [31] |
| Aroma | Clove [41] |
| Texture | Not available [41] |
| Color | Yellow [28,31] |
| Acidity | 0.77 [41] |
| pH | 2.6–3.33 [31]<br>3.96 [18,24] |

*4.4. Antinutritional and Toxicological Properties of Strychnos spinosa*

Antinutritional properties have an adverse effect on the food digestion in the light of the food classes, such as protein and carbohydrates, and decrease the bioavailability of minerals, such as iron and zinc [49,54,55]. The reported components of such in *Strychnos spinosa* were low and below the established toxic level [56,57]. The seeds of *Strychnos spinosa* contain strychnine and are bitter tasting [31,58]. Toxic alkaloids are present in the seeds and unripe pulp of *Strychnos spinosa* [58].

**5. Postharvest Handling, Preservation, Storage, and Processing of *Strychnos spinosa***

*5.1. Postharvest Handling*

*Strychnos spinosa* fruits are harvested by shaking, hitting, knocking, or plucking the trees [18]. On the other hand, unripe *Strychnos spinosa* fruits are harvested and buried under a light sand for months, until it is ripe, in order to prevent postharvest losses [16,26,59]. The fruit pulp usually changes from its dry texture to a golden color after storage and, hence, is ready to be consumed [18]. As applicable with other climacteric fruits, during storage, *Strychnos spinosa* increase in soluble solid content and accumulate glucose, sucrose, and fructose [26]. The slow spoilage attributed to the fruit can be linked to the hard texture that assists in resisting insects and pathogens [16,60–62].

### 5.2. Products Preservation

*Strychnos spinosa* can be processed to dried products, but the preparation methods and conditions vary across locations in a small-scale level. Postharvest processing of *Strychnos spinosa* can be achieved through drying, juicing, maceration, and cooking. Although, storage influences the bioavailability and physical characteristics of the plant [63]. In southern Africa, *Strychnos spinosa* fruits are often dried by fire and or direct sunlight too, and thereafter grinded into flour [18]. Additionally, the sun-dried *Strychnos spinosa* pulp can be kept for 2 months to 5 years, making heat-drying a good preservation method for the rural communities [64]. The moisture content of *Strychnos spinosa* fruit ranges from 60 to 91%, which mainly depends on the degree and method of heating [56,65]. In addition, a properly dried fruit product does have a residual moisture content that ranges between 18 and 24%, with a good shelf-life [66,67].

### 5.3. Advantages and Challenges of Processing Techniques

Currently, the impact of processing *Strychnos spinosa* and the assessment of its contribution to nutrient uptake is not well documented. Therefore, optimization of the processing and profiling of the food value of *Strychnos spinosa* and its products is important for the improvement of the processing procedures, which has the potential to increase the demand for the plant and its products (Figure 3). On this basis, we have identified several advantages, disadvantages, and recommendations for processing *Strychnos spinosa*. Considering the nutritional quality of the fruit, it may easily serve as an important source of nutrients for children and pregnant women [17,54,68]. Thus, improved processing of *Strychnos spinosa* fruit could be a sustainable solution to the problems of the rural communities [18].

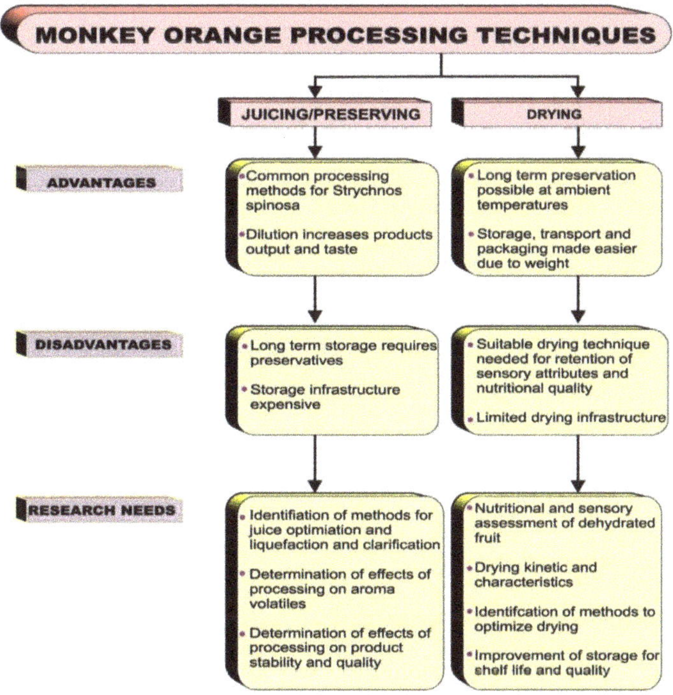


**Figure 3.** Products, processing, and way forward for *Strychnos spinosa*.

*5.4. Nutritional Quality and Economic Potential of Strychnos spinosa*

*Strychnos spinosa* fruit and its byproducts can contribute to the economy and rural livelihood in Africa. This undervalued plant has potential that can make it withstand market competition with respect to exotic fruits (e.g., orange and strawberry). The high nutritional components and diverse phytochemicals in the plant confer immense benefits. Hence, large-scale production, marketing, and trading of *Strychnos spinosa* fruit remain important for sustainable livelihood and economic development, especially in the rural communities. Presently, there is paucity of knowledge, with limited literature on the several aspects of the fruit [69]. The commercialization of *Strychnos spinosa* will remain low until the economic returns on investment associated with the domestication of the fruit tree are profitable [70].

## 6. Domestication of *Strychnos spinosa*, Cultivation Problems, and Way Forward

*Strychnos spinosa* has been cultivated in southern Africa but without tangible results [9,71]. To date, no trials of the cultivation of *Strychnos spinosa* have been conducted in Africa; hence, the fruit tree is mainly sourced from the wild populations. The problems experienced by the rural populations concerning the cultivation of the underutilized fruit as a crop are: (1) land available, (2) slow growth cycle, minimal yield, and (3) common fast-cash economic culture [17,72]. Enhanced and effective information dissemination, including findings and activities, may improve as more stakeholders participate (Figure 4).

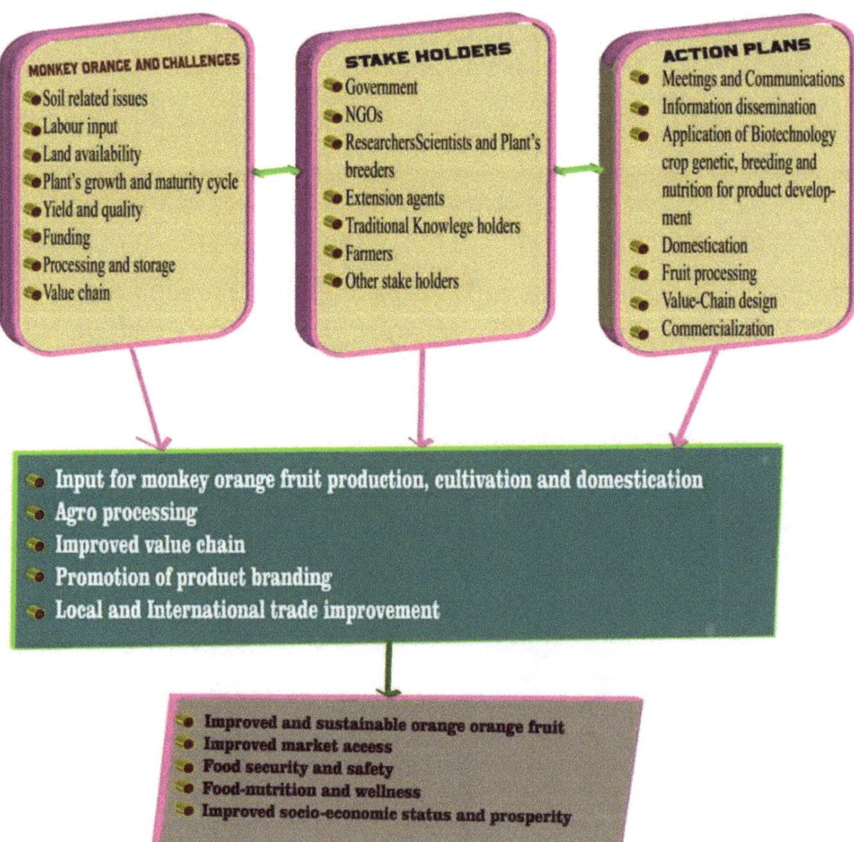


**Figure 4.** Schematic framework of priority areas for intervention on *Strychnos spinosa*.

There is need for active and effective collaborations by the stakeholders on *Strychnos spinosa*. Research findings on the plants can be disseminated to the rural communities, through local NGOs and other relevant stakeholders, such as the agricultural extension services. Improving processing of *Strychnos spinosa* can enhance the possibilities for its domestication, agro-processing, production, and commercialization [29]. These envisaged findings will be useful to *Strychnos spinosa* and the much-needed intervention in research of indigenous fruit trees.

*Areas for Further Research*

Sensory and nutritional composition of *Strychnos spinosa* during storage is not available. There is paucity of information on the suitability of the drying methods for *Strychnos spinosa*. Therefore, further studies on the suitability of dried products and characteristics need to be conducted to establish a drying method that fits local conditions and the possibility for commercialization. Furthermore, few studies have evaluated the nutritional and sensorial characteristics of fresh *Strychnos spinosa* juice [18]. Therefore, improving the production processes of *Strychnos spinosa* through preservation technique optimization needs to be investigated. Exploration of the value chain to enhance the economic value and potential of *Strychnos spinosa* is needed. Finally, research by the plant scientists and breeders on the domestication of *Strychnos spinosa* needs to be given more priority, owing to its commercial, nutritional, and economic potential.

## 7. Conclusions and Recommendations

*Strychnos spinosa* fruit have the potential to impart livelihood benefits and improve the nutritional status, as well as the economic prosperity, of the rural population. The micronutrients and macronutrients in the fruit tree are key to its relevance. On this basis, *Strychnos spinosa* is an important food source for children, pregnant women, and the poor. Nonetheless, limited research has been conducted regarding the value addition and processing for *Strychnos spinosa* in comparison with many popular and commercial fruits. The plant has great potential in the African rural communities, since the local environmental conditions are appropriate for its cultivation. In order to mitigate some of the existing challenges affecting the domestication of the plant for commercialization, there is need for trans-disciplinary research by different stakeholders, as well as the suggested action plan to improve the problems associated with the cultivation of the plant. Overall, we proposed priority areas for policy and intervention, and recommend an all-inclusive and sustainable development approach, as *Strychnos spinosa* could contribute to the attainment of the food-nutrition target of the United Nations Sustainable Development Goals (UN SDG, 2030).

**Author Contributions:** Conceptualization, A.O.O. and A.O.A.; methodology, A.O.O. and A.O.A.; writing—original draft preparation, A.O.O.; writing—review and editing, A.O.A. All authors have read and agreed to the published version of the manuscript.

**Funding:** Adeyemi Oladapo Aremu received research fund from the National Research Foundation NRF, Incentive Funding for Rated Researchers, Grant UID: 109508. We appreciate the institutional support (including the payment of the APC) from the Food Security and Safety Niche Area, the Faculty of Natural and Agricultural Sciences (FNAS), North-West University, Mmabatho, South Africa.

**Acknowledgments:** Moloko Mojapelo's (Department of Agriculture, Forestry and Fisheries Directorate Plant Production: Indigenous Food Crops Division) support in providing the photographs of *Strychnos spinosa* used in this review is appreciated.

**Conflicts of Interest:** The authors declare no conflict of interest.

# References


1. Emadi, M.H.; Rahmanian, M. Commentary on Challenges to Taking a Food Systems Approach within the Food and Agriculture Organization (FAO). In *Food Security and Land Use Change under Conditions of Climatic Variability*; Springer: Cham, Switzerland, 2020; pp. 19–31.
2. Food and Agriculture Organization (FAO); International Fund for Agricultural Development(IFAD); United Nations Children's Fund (UNICEF); World Food Program(WFP); World Health Organization (WHO). *The State of Food Security and Nutrition in the World 2020*; Food and Agriculture Organization: Rome, Italy, 2020.
3. Gomez, M.I. *A Resource Inventory of Indigenous and Traditional Foods in Zimbabwe*; University of Zimbabwe Publications: Zambezia, Mozambique, 1989.
4. Omotayo, A.O. Farming households' environment, nutrition and health interplay in Southwest, Nigeria. *Int. J. Sci. Res. Agric. Sci.* **2016**, *3*, 84–98. [CrossRef]
5. Omotayo, A.O.; Aremu, B.R.; Alamu, O.P. Food utilization, nutrition, health and farming households' income: A critical review of literature. *J. Hum. Ecol.* **2016**, *56*, 171–182. [CrossRef]
6. Omotayo, A.O. Parametric assessment of household's food intake, agricultural practices and health in rural South West, Nigeria. *Heliyon* **2020**, *6*, e05433. [CrossRef] [PubMed]
7. Omotoso, A.B.; Daud, A.S.; Adebayo, R.A.; Omotayo, A.O. Socioeconomic determinants of rural households' food crop production in Ogun state, Nigeria. *Appl. Ecol. Environ. Res.* **2018**, *16*, 3627–3635. [CrossRef]
8. Laverdiere, M.; Mateke, S. Non-Wood Forest Products: Food for Life. Indigenous Fruit Trees in Southern Africa. Available online: https://agris.fao.org/agris-search/search.do?recordID=XF2015045277 (accessed on 22 May 2021).
9. Leakey, R.R.B.; Akinnifesi, F.K. Towards a Domestication Strategy for Indigenous Fruit Trees in the Tropics. In *Indigenous Fruit Trees in the Tropics: Domestication, Utilization and Commercialization*; CABI: Wallingford, UK, 2008; pp. 28–49.
10. Van Wyk, B.E. The potential of South African plants in the development of new food and beverage products. *S. Afr. J. Bot.* **2011**, *77*, 857–868. [CrossRef]
11. Cemansky, R. Africa's indigenous fruit trees: A blessing in decline. *Environ. Health Perspect.* **2015**, *123*, A291–A296. [CrossRef] [PubMed]
12. Richmond, R.; Bowyer, M.; Vuong, Q. Australian native fruits: Potential uses as functional food ingredients. *J. Funct. Foods* **2019**, *62*, 103547. [CrossRef]
13. Pfukwa, T.M.; Chikwanha, O.C.; Katiyatiya, C.L.F.; Fawole, O.A.; Manley, M.; Mapiye, C. Southern African indigenous fruits and their byproducts: Prospects as food antioxidants. *J. Funct. Foods* **2020**, *75*, 104220. [CrossRef]
14. Omotayo, A.O.; Ndhlovu, P.T.; Tshwene, S.C.; Aremu, A.O. Utilization pattern of indigenous and naturalized plants among some selected rural households of North West province, South Africa. *Plants* **2020**, *9*, 953. [CrossRef]
15. Madzimure, J.; Nyahangare, E.T.; Hamudikuwanda, H.; Hove, T.; Belmain, S.R.; Stevenson, P.C.; Mvumi, B.M. Efficacy of *Strychnos spinosa* (Lam.) and *Solanum incanum* L. aqueous fruit extracts against cattle ticks. *Trop. Anim. Health Prod.* **2013**, *45*, 1341–1347. [CrossRef]
16. Mwamba, C.K. *Monkey Orange: Strychnos Cocculoides*; Crops for the Future: Southampton, UK, 2006.
17. Omotayo, A.O.; Aremu, A.O. Underutilized African indigenous fruit trees and food–nutrition security: Opportunities, challenges, and prospects. *Food Energy Secur.* **2020**, *9*, 220. [CrossRef]
18. Ngadze, R.T.; Linnemann, A.R.; Nyanga, L.K.; Fogliano, V.; Verkerk, R. Local processing and nutritional composition of indigenous fruits: The case of monkey orange (*Strychnos* spp.) from Southern Africa. *Food Rev. Int.* **2017**, *33*, 123–142. [CrossRef]
19. Hedberg, I.; Hedberg, O.; Madat, P.J.; Mshigeni, K.E.; Mshiu, E.; Samuelsson, G. Inventory of plants used in traditional medicine in Tanzania. II. Plants of the families dilleniaceae—*Opiliaceae*. *J. Ethnopharmacol.* **1983**, *9*, 105–127. [CrossRef]
20. Kokwaro, J. Medicinal plants of East Africa. East African Literature Bureau. University of Nairobi Press: Nairobi, Kenya 1976, 45. Available online: http://41.204.161.209/handle/11295/34820 (accessed on 1 December 2021).
21. Aremu, A.O.; Moyo, M. Health Benefits and Biological Activities of Spiny Monkey Orange (*Strychnos spinosa* Lam.): An African Indigenous Fruit Tree. *J. Ethnopharmacol.* **2022**, *283*, 114704. [CrossRef] [PubMed]
22. Anjarwalla, P.; Belmain, S.; Sola, P.; Jamnadass, R.; Stevenson, P.C. Guide des Plantes Pesticides. In *Optimisation des Plantes Pesticides: Technologie, Innovation, Sensibilisation et Reseaux*; World Agroforestry Centre: Nairobi, Kenya, 2016.
23. Guissou, K.M.L.; Kristiansen, T.; Lykke, A.M. Local perceptions of food plants in eastern Burkina Faso. *Ethnobot. Res. Appl.* **2015**, *14*, 199–209. [CrossRef]
24. Ngadze, R.T. Value Addition of Southern African Monkey Orange (*Strychnos* spp.): Composition, Utilization and Quality. Ph.D. Thesis, Wageningen University, Wageningen, The Netherlands, October 2018.
25. Shai, K.N.; Ncama, K.; Ndhlovu, P.T.; Struwig, M.; Aremu, A.O. An exploratory study on the diverse uses and benefits of locally-sourced fruit species in three villages of Mpumalanga Province, South Africa. *Foods* **2020**, *9*, 1581. [CrossRef] [PubMed]
26. Sitrit, Y.; Loison, S.; Ninio, R.; Dishon, E.; Bar, E.; Lewinsohn, A.E.; Mizrahi, Y. Characterization of monkey orange (*Strychnos spinosa* Lam.), a potential new crop for arid regions. *J. Agric. Food Chem.* **2003**, *51*, 6256–6260. [CrossRef]
27. Avakoudjo, H.G.G.; Hounkpèvi, A.; Idohou, R.; Koné, M.W.; Assogbadjo, A.E. Local knowledge, uses, and factors determining the use of *Strychnos spinosa* organs in Benin (West Africa). *Econ. Bot.* **2020**, *74*, 15–31. [CrossRef]

28. Thiombiano, D.N.E.; Lamien, N.; Castro-Euler, A.M.; Vinceti, B.; Agundez, D.; Boussim, I.J. Local communities demand for food tree species and the potentialities of their landscapes in two ecological zones of Burkina Faso. *Open J. For.* **2013**, *03*, 79–87. [CrossRef]

29. Omotayo, A.O.; Ijatuyi, E.J.; Ogunniyi, A.I.; Aremu, A.O. Exploring the resource value of Transvaal red milk wood (*Mimusops zeyheri*) for food security and sustainability: An appraisal of existing evidence. *Plants* **2020**, *9*, 1486. [CrossRef]

30. White, R.A., III; Chan, A.M.; Gavelis, G.S.; Leander, B.S.; Brady, A.L.; Slater, G.F.; Lim, D.S.S.; Suttle, C.A. Metagenomic analysis suggests modern freshwater microbialites harbor a distinct core microbial community. *Front. Microbiol.* **2016**, *6*, 1531. [CrossRef]

31. Nitcheu Ngemakwe, P.N.; Remize, F.; Thaoge, M.; Sivakumar, D. Phytochemical and nutritional properties of underutilised fruits in the southern African region. *S. Afr. J. Bot.* **2017**, *113*, 137–149. [CrossRef]

32. Van Wyk, B.-E.; Gericke, N. *People's Plants: A Guide to Useful Plants of Southern Africa*; Briza Publications: Pretoria, South Africa, 2000.

33. Rodrigues, S.; de Brito, E.S.; de Oliveira Silva, E. Maboque/Monkey Orange—*Strychnos spinosa*. In *Exotic Fruits*; Rodrigues, S., de Oliveira Silva, E., de Brito, E.S., Eds.; Elsevier: Amsterdam, The Netherlands, 2018; pp. 293–296.

34. Van Wyk, B. *Field Guide to Trees of Southern Africa*; Penguin Random House South Africa: Cape Town, South Africa, 2013.

35. Mizrahi, Y.; Nerd, A.; Sitrit, Y. New Fruits for Arid Climates. In *Trends in New Crops and New Uses*; ASHS Press: Alexandria, VA, USA, 2002; pp. 378–384.

36. Amarteifio, J.O.; Mosase, M.O. The chemical composition of selected indigenous fruits of Botswana. *J. Appl. Sci. Environ. Manag.* **2006**, *10*, 43–47. [CrossRef]

37. Lawal, F.; Bapela, M.J.; Adebayo, S.A.; Nkadimeng, S.M.; Yusuf, A.A.; Malterud, K.E.; McGaw, L.J.; Tshikalange, T.E. Anti-inflammatory potential of South African medicinal plants used for the treatment of sexually transmitted infections. *S. Afr. J. Bot.* **2019**, *125*, 62–71. [CrossRef]

38. Shajib, M.T.I.; Kawser, M.; Nuruddin Miah, M.; Begum, P.; Bhattacharjee, L.; Hossain, A.; Fomsgaard, I.S.; Islam, S.N. Nutritional composition of minor indigenous fruits: Cheapest nutritional source for the rural people of Bangladesh. *Food Chem.* **2013**, *140*, 466–470. [CrossRef] [PubMed]

39. Lockett, C.T.; Calvert, C.C.; Grivetti, L.E. Energy and micronutrient composition of dietary and medicinal wild plants consumed during drought. Study of rural Fulani, Northeastern Nigeria. *Int. J. Food Sci. Nutr.* **2000**, *51*, 195–208. [CrossRef] [PubMed]

40. Nile, S.H.; Park, S.W. Edible berries: Bioactive components and their effect on human health. *Nutrition* **2014**, *30*, 134–144. [CrossRef] [PubMed]

41. Alothman, M.; Bhat, R.; Karim, A. Effects of radiation processing on phytochemicals and antioxidants in plant produce. *Trends Food Sci. Technol.* **2009**, *20*, 201–212. [CrossRef]

42. Hur, S.J.; Lee, S.Y.; Kim, Y.-C.; Choi, I.; Kim, G.-B. Effect of fermentation on the antioxidant activity in plant-based foods. *Food Chem.* **2014**, *160*, 346–356. [CrossRef] [PubMed]

43. Shahidi, F.; Ambigaipalan, P. Phenolics and polyphenolics in foods, beverages and spices: Antioxidant activity and health effects–A review. *J. Funct. Foods* **2015**, *18*, 820–897. [CrossRef]

44. Alasalvar, C.; Shahidi, F. Composition, phytochemicals, and beneficial health effects of dried fruits: An overview. *Dried Fruits Phytochem. Health Eff.* **2013**, 1–18. [CrossRef]

45. Shofian, N.M.; Hamid, A.A.; Osman, A.; Saari, N.; Anwar, F.; Pak Dek, M.S.; Hairuddin, M.R. Effect of freeze-drying on the antioxidant compounds and antioxidant activity of selected tropical fruits. *Int. J. Mol. Sci.* **2011**, *12*, 4678–4692. [CrossRef] [PubMed]

46. Brielmann, H.L.; Setzer, W.N.; Kaufman, P.B.; Kirakosyan, A.; Cseke, L.J. Phytochemicals: The chemical components of plants. In *Natural Products from Plants*, 2nd ed.; Cseke, L.J., Kirakosyan, A., Kaufman, P.B., Warber, S.L., Duke, J.A., Brielmann, H.L., Eds.; Taylor and Francis Group: Boca Raton, FL, USA, 2006; Volume 2, pp. 1–49.

47. Adinortey, M.B.; Sarfo, J.K.; Adinortey, C.A.; Ofori, E.G.; Kwarteng, J.; Afrifa, J.; Owusu, F.A. Inhibitory effects of *Launaea taraxacifolia* and *Strychnos spinosa* leaves extract on an isolated digestive enzyme linked to type -2 -diabetes mellitus. *J. Biol. Life Sci.* **2018**, *9*, 52–65. [CrossRef]

48. Garba, A.; Muhammad, A.; Ahmad, K.A. In vitro antitrypanosomal and antioxidant activities of diethyl pthalate from the stem bark of *Strychnos spinosa* Lam. *Int. J. Med. Biosci.* **2018**, *1*, 1–22.

49. Kumar, Y.; Basu, S.; Goswami, D.; Devi, M.; Shivhare, U.S.; Vishwakarma, R.K. Anti-nutritional compounds in pulses: Implications and alleviation methods. *Legume Sci.* **2021**, *e111*, 1–13. [CrossRef]

50. Ahmad, G.; Muhammad, A.; Ahmad, K.; Gbeminiyi, A. Phytochemical and antimicrobial studies of the root bark extracts of *Strychnos spinosa* Lam. *Bayero J. Pure Appl. Sci.* **2020**, *12*, 161–165. [CrossRef]

51. Ugoh, S.C.; Bejide, O.S. Phytochemical screening and antimicrobial properties of the leaf and stem bark extracts of *Strychnos spinosa*. *Nat. Sci.* **2013**, *11*, 123–128.

52. Nwozo, S.O.; Ajayi, I.A.; Obadare, M. Phytochemical screening and antimicrobial activity of 10 medicinal seeds from Nigeria. *Med. Aromat. Plant Sci. Biotechnol.* **2010**, 73–75. Available online: http://www.globalsciencebooks.info/Online/GSBOnline/images/2010/MAPSB_4(1)/MAPSB_4(1)73-75o.pdf (accessed on 1 December 2021).

53. Tittikpina, N.K.; Atakpama, W.; Hoekou, Y.; Diop, Y.M.; Batawila, K.; Akapagana, K. *Strychnos spinosa* Lam: Comprehensive review on its medicinal and nutritional uses. *Afr. J. Tradit. Complement. Altern. Med.* **2020**, *17*, 8–28. [CrossRef]

54. Bille, P.; Shikongo-Nambabi, M.; Cheikhyoussef, A. Value addition and processed products of three indigenous fruits in Namibia. *Afr. J. Food Agric. Nutr. Dev.* **2013**, *13*, 7192–7212. [CrossRef]
55. Soetan, K.; Olaiya, C.O.; Oyewole, O.E. The importance of mineral elements for humans, domestic animals and plants—A review. *Afr. J. Food Sci.* **2010**, *4*, 200–222.
56. Bello, M.; Falade, O.; Adewusi, S.; Olawore, N. Studies on the chemical compositions and anti-nutrients of some lesser known Nigeria fruits. *Afr. J. Biotechnol.* **2008**, *21*, 7.
57. Msonthi, J.; Galeffi, C.; Nicoletti, M.; Messana, I.; Marini-Bettolo, G. Kingiside aglucone, a natural secoiridoid from unripe fruits of *Strychnos spinosa*. *Phytochemistry* **1985**, *24*, 771–772. [CrossRef]
58. Nyahangare, E.T.; Hove, T.; Mvumi, B.M.; Hamudikuw, H.; Belmain, S.R.; Madzimure, J.; Stevenson, P.C. Acute mammalian toxicity of four pesticidal plants. *J. Med. Plants Res.* **2012**, *6*, 2674–2680.
59. Rampedi, I.T.; Olivier, J. Traditional beverages derived from wild food plant species in the Vhembe District, Limpopo Province in South Africa. *Ecol. Food Nutr.* **2013**, *52*, 203–222. [CrossRef] [PubMed]
60. Mkonda, A.; Akinnifesi, F.; Swai, R.; Kadzere, I.; Kwesiga, F.; Maghembe, J.; Saka, J.; Lungu, S.; Mhango, J. Towards domestication of 'wild orange' *Strychnos cocculoides* in Southern Africa: A synthesis of research and development efforts. In Proceedings of the Regional Agroforestry Conference on Agroforestry Impacts on Livelihoods in Southern Africa: Putting research into Practice, Warmbaths, South Africa, 20–24 May 2002; World Agroforestry Centre (ICRAF): Nairobi, Kenya, 2002; pp. 67–76.
61. Ruffo, C.; Birnie, A.; Tengnas, B. Edible Wild Plants of Tanzania (Volume 27). Regional Land Management Unit/Sida. Available online: http://apps.worldagroforestry.org/downloads/Publications/PDFS/B11913.pdf (accessed on 1 December 2021).
62. Saka, J.; Swai, R.; Mkonda, A.; Schomburg, A.; Kwesiga, F.; Akinnifesi, F. Processing and utilisation of indigenous fruits of the miombo in southern Africa. In Proceedings of the Agroforestry Impacts on Livelihoods in Southern Africa: Putting Research into Practice, Warmbaths, South Africa; 2002; pp. 20–24. Available online: https://www.worldagroforestry.org/publication/processing-and-utilisation-indigenous-fruits-miombo-southern-africa (accessed on 1 December 2021).
63. Chong, C.H.; Law, C.L.; Figiel, A.; Wojdyło, A.; Oziembłowski, M. Colour, phenolic content and antioxidant capacity of some fruits dehydrated by a combination of different methods. *Food Chem.* **2013**, *141*, 3889–3896. [CrossRef] [PubMed]
64. Shackleton, C.; Dzerefos, C.; Shackleton, S.; Mathabela, F. The use of and trade in indigenous edible fruits in the Bushbuckridge savanna region, South Africa. *Ecol. Food Nutr.* **2000**, *39*, 225–245. [CrossRef]
65. Malaisse, F.; Parent, G. Edible wild vegetable products in the Zambezian woodland area: A nutritional and ecological approach. *Ecol. Food Nutr.* **1985**, *18*, 43–82. [CrossRef]
66. Pátkai, G. Fruit and Fruit Products as Ingredients. In *Handbook of Fruits and Fruit Processing*; Wiley: California, CA, USA, 2012; pp. 263–275. [CrossRef]
67. Nicoli, M.; Anese, M.; Parpinel, M. Influence of processing on the antioxidant properties of fruit and vegetables. *Trends Food Sci. Technol.* **1999**, *10*, 94–100. [CrossRef]
68. Legwaila, G.; Mojeremane, W.; Madisa, M.; Mmolotsi, R.; Rampart, M. Potential of traditional food plants in rural household food security in Botswana. *J. Hortic. For.* **2011**, *3*, 171–177.
69. Ham, C.; Akinnifesi, F.K.; Franzel, S.; Jordaan, D.; Hansmann, C.; Ajayi, O.; De Kock, C. Opportunities for Commercialization and Enterprise Development of Indigenous Fruits in Southern Africa. In *Indigenous Fruit Trees in the Tropics: Domestication, Utilization and Commercialization*; CABI: Wallingford, UK, 2008; pp. 254–272.
70. Akinnifesi, F.; Ajayi, O.; Sileshi, G.; Kadzere, I.; Akinnifesi, A.I. Domesticating and Commercializing Indigenous Fruit and Nut Tree Crops for Food Security and Income Generation in Sub-Saharan Africa. Available online: https://www.worldagroforestry.org/publication/domesticating-and-commercializing-indigenous-fruit-and-nut-tree-crops-food-security-and (accessed on 1 December 2021).
71. Thiong'O, M.; Kingori, S.; Jaenicke, H. The taste of the wild: Variation in the nutritional quality of marula fruits and opportunities for domestication. In Proceedings of the International Symposium on Tropical and Subtropi-cal Fruits 575, Cairns, Australia, 25 November–1 December 2000; pp. 237–244. [CrossRef]
72. Jackson, J.C.; Duodu, K.G.; Holse, M.; de Faria, M.D.L.; Jordaan, D.; Chingwaru, W.; Hansen, A.; Cencic, A.; Kandawa-Schultz, M.; Mpotokwane, S.M.; et al. The Morama Bean (*Tylosema Esculentum*): A Potential Crop for Southern Africa. In *Advances in Food and Nutrition Research*; Elsevier: Amsterdam, The Netherlands, 2010; Volume 61, pp. 187–246.

*Article*

# Nutritional Characteristics Assessment of Sunflower Seeds, Oil and Cake. Perspective of Using Sunflower Oilcakes as a Functional Ingredient

Ancuţa Petraru , Florin Ursachi and Sonia Amariei *

Faculty of Food Engineering, Stefan cel Mare University of Suceava, 720229 Suceava, Romania; ancuta.petraru@fia.usv.ro (A.P.); florin.ursachi@fia.usv.ro (F.U.)
* Correspondence: sonia@usm.ro; Tel.: +40-740-311-291

**Abstract:** Ample amounts of by-products are generated from the oil industry. Among them, sunflower oilcakes have the potential to be used for human consumption, thus achieving the concept of sustainability and circular economy. The study assessed the nutritional composition of sunflower seeds, cold-pressed oil and the remaining press-cakes with the aim of its valorization as a food ingredient. Sunflower oil contains principally oleic (19.81%) and linoleic (64.35%) acids, which cannot be synthetized by humans and need to be assimilated through a diet. Sunflower seeds are very nutritive (33.85% proteins and 65.42% lipids and 18 mineral elements). Due to the rich content of lipids, they are principally used as a source of vegetable oil. Compared to seeds, sunflower oilcakes are richer in fibers (31.88% and 12.64% for samples in form of pellets and cake, respectively) and proteins (20.15% and 21.60%), with a balanced amino acids profile. The remaining oil (15.77% and 14.16%) is abundant in unsaturated fatty acids (95.59% and 92.12%). The comparison between the three products showed the presence of valuable components that makes them suitable for healthy diets with an adequate intake of nutrients and other bioactive compounds with benefic effects.

**Keywords:** sustainability; sunflower seeds; sunflower oil; sunflower oilcakes; nutritive parameters; classification; amino acids profile; fatty acids composition

**Citation:** Petraru, A.; Ursachi, F.; Amariei, S. Nutritional Characteristics Assessment of Sunflower Seeds, Oil and Cake. Perspective of Using Sunflower Oilcakes as a Functional Ingredient. *Plants* **2021**, *10*, 2487. https://doi.org/10.3390/plants10112487

Academic Editor: Ivo Vaz de Oliveira

Received: 23 October 2021
Accepted: 16 November 2021
Published: 17 November 2021

**Publisher's Note:** MDPI stays neutral with regard to jurisdictional claims in published maps and institutional affiliations.

## 1. Introduction

Large amounts of biodegradable waste and residues are produced and discarded every year from the food industry. These residues have high biochemical and chemical oxygen demand. For this reason, untreated waste harms the microflora [1]. Food waste with a high fat content is susceptive to oxidation and thus harmful for microflora due to continuous enzymatic activity, which accelerates spoilage and limits technological possibilities of disposal. Environmental protection must be the first priority of international politics [2]. Nowadays, due to the rapid expansion of the human population and current environmental issues (over-exploitation of resources, degradation of environment), it is necessary for a transition to the circular economy and the development of new strategies to minimize food waste during the supply chain [3,4]. Moreover, problems related to cost disposal and the use of by-products have become an increasing challenge [5]. Conventional methods for waste disposal and valorization are incineration, aerobic fermentation, composting, fertilizer and feedstuff [6]. Alternative solutions consist in extracting the maximum value from waste, hence the bioactive compound and re-circulating them in the process, making value-added products and thus creating the concept of "waste = food" [7,8].

Oilseeds are mostly used as a source of vegetable oils. After the extraction process, large amounts of residues and by-products are available. The use of these permits the achievement of effective waste utilization and the successful realization of the circular economy concept [9]. Possible valorization methods of the oilcakes involve their use in animal diets as feeds, compost, a substrate in the production of enzymes, antibiotics and biosurfactants and in the recovery of bioactive compounds for further use in the production of new value-added products [6,10].

Sunflower is a plant from the *Asteraceae* family, *Helianthus* genus and more than seventy species are known worldwide. The origin of the name derived from the aspect of the plant which resembled a sun and the fact that it rotates after the sun's rays [11,12]. Globally, sunflower seeds are ranked as one of the most produced oilseeds crops alongside rapeseed, soybean and cottonseed [13]. Their composition and nutritive values depend on numerous factors, namely genotype, soil type, agricultural practices, climatic and processing conditions [14]. Two types of sunflower seeds are known, namely the oil-producing seed and the ones used for confectionary purposes. The first is black with a thin hull (lignin and cellulolytic materials) that adheres to the kernel and represents 20% of the total weight. Originally, the seeds contained 25% oil but by modern plant breeding methods [15,16] (induced mutation, hybridization, molecular breeding) new sunflower hybrids, in which the oil content was increased to 40% [11,17,18], were created. The seeds are a source of dietary fibers, unsaturated fatty acids (more linoleic than oleic), antioxidants, flavonoids (quercetin, luteolin, apigenin and kaempferol) amino acids, proteins (up to 20%), vitamins (E, B, folate and niacin) and minerals (principally calcium, copper, iron, magnesium, manganese, selenium, phosphorous, potassium, sodium and zinc) [19,20]. The amino acids profile includes glutamic, aspartic acids, arginine, phenylalanine, tyrosine, leucine, methionine and cysteine. The content in fatty acids varies up to 31%, being higher than the other oilseed such as safflower, peanut, soybean, sesame and flaxseed [21].

The oilseeds are mostly used as a source of vegetable oil with unique physicochemical properties [22]. The traditional extraction techniques involve the use of a mechanic press (hot or cold pressing) or solvents [23,24]. Sunflower oil, due to its easy accessibility and numerous health benefits (maintenance of low cholesterol and low-density lipoprotein levels in the human body, antioxidant, anticancer, antihypertensive, anti-inflammatory, skin protective and analgesic), is widely preferred in Europe, Mexico and several countries of South America. After extraction, it remains liquid at room temperature and has a shelf-life of over one year at 10 °C and in darkness [11,25]. The major components are linoleic (59–65%) and oleic acids (30–70%). These represent 48−78% of the total fatty acids profile. There is also a small percentage of palmitic and stearic acids (15% for both fatty acids) present [26,27]. Sunflower oil is also rich in vitamins (important role in the good functioning of the skin, nerves and digestive system), minerals (role in the enzymatic and metabolic processes) and excellent phytochemical such as carotenoids, tocopherols, phenols and tocotrienols with antioxidant activity. The variation of the composition depends on the plant's species and the extraction methods employed [28,29].

Oilcakes are the principal by-products obtained after the extraction of oil from the seeds. Then they are air-dried to remove the water before storage. Sometimes they are molded into two forms, namely flour (ground material) and pellet. The term is synonymous with press-cake, meal and oil meal. In terms of appearance, sunflower meal has the taste and the smell characteristic of the initial raw material without musty, mold, rancid and foreign smells. The color changes from black to gray [30].

Generally, the meal is used in animal diet as feeds because it is an excellent source of protein and thus produces an increase in biomass [31–33]. It can be also used for human consumption. The essential amino acids present in sunflower press-cakes are cysteine, methionine, leucine, valine, isoleucine, tryptophan, alanine and phenylalanine [25]. Regarding the minerals and vitamins, phosphorus, thiamine, nicotinic, pantothenic acids and riboflavin are predominant [26]. The dehulled process reduces the fiber content and increases the protein content [34]. The difference in chemical composition varies depending on variety, growing condition, dehulling and extraction method [35].

The physical characteristics are extremely important in the seeds production chain for designing various agricultural machines and equipment for operations such as planting, harvesting, cleaning, quality assessment, classification, dehulling, milling, packaging, storing and oil extraction [36]. Physical characteristics can be grouped into four categories, namely dimensional, gravimetric, compressive and frictional (angle of response and static friction) [37,38]. Length, width, thickness, area, volume, sphericity, equivalent diameter and projected area are dimensional proprieties and offer information about the shape. Instead, mass, bulk density, true density and porosity are gravimetric properties [39]. The size, shape and density influence aerodynamic proprieties, which are crucial in designing a harvester [40]. Moreover, hulling efficiency is affected by the hull structure, seed size and density. Sphericity, static friction and angle of response play a key role in designing storage facilities, while porosity in designing extraction machinery [41]. When evaluating quality, consumers choose their preference based on texture, flavor and appearance. In order to reduce damaged and defective seeds, a classifier that complies with the quality indicators found should be realized [42,43].

The aim of this study was to investigate the physicochemical properties of seeds, oil and oilcakes and the transformations that take place in the processing stages of the raw material. This study is important to evaluate the losses in the processing and to discover the bioactive compounds that can be extracted from the by-products after oil extraction. This study was also conducted to determine the physical attributes of seeds and kernels in order to classify them into quality classes.

## 2. Results and Discussions

### 2.1. Chemical Composition of the Seeds, Kernels and Hulls

Sunflower seeds consist of kernels and seed coats or hulls. Sunflower hulls represent 21–30% of the seed weight and generally are considered a waste by-product [44]. From the hulls can be recovered valuable phenolic compound and cellulose fibers for the production of green, renewable, biodegradable and edible food packaging material, thus reducing the global plastic production [45]. Another alternative to dispose of them is by transforming the biomass (by pyrolysis, gasification and fermentation) obtaining bio-oils rich in furfural content (a valuable bio-renewable chemical that can be used in the production of biofuels and biochemicals) [46] In our study, the proportion of hulls calculated for the sunflower seeds fell into the range from 13.67% up to 43.47%. Values depend on variety, environmental conditions, seed size and oil content [47]. The lower values can be the result of the continuous effort to increase the seed's oil content [40]. The hulls contain a low percentage of proteins (7.82%), lipids (8.81%) and ash (2.45%) according to Table 1. Similar values were found by other authors [40,46,48], values ranged within 3.48–12.40% for moisture, 2.3–9.45% for oil, 5.36–7.36% for protein and 2.1–4.11% for ash.

**Table 1.** Chemical composition of the whole seeds, kernels and hulls.

| Sample | Moisture, % | Ash, % | Proteins, % | Lipids, % |
|---|---|---|---|---|
| Seed | 6.16 ± 0.04 [b] | 2.73 ± 0.04 [b] | 33.85 ± 0.88 [b] | 65.42 ± 0.4 [a] |
| Kernel | 4.60 ± 0.03 [c] | 3.31 ± 0.11 [a] | 23.73 ± 1.31 [a] | 32.50 ± 2.21 [b] |
| Hull | 7.88 ± 0.09 [a] | 2.45 ± 0.11 [c] | 7.82 ± 0.22 [c] | 8.81 ± 0.12 [c] |

Different superscripts letters after the values indicated differences statistically significant at $p < 0.05$%.

Nutritional characteristics of mature and sun-dried sunflower seeds, hulls and kernels are summarized in Table 1. The findings showed that seeds contain on average 33.85% proteins, 65.42% lipids and 2.73% ash, most of which are found in the kernels.

In comparison with other species, the seeds are rich in crude fat, due probably to the breeding techniques (induced mutation, hybridization, molecular breeding) used to increase the oil content. Various studies about the chemical composition of high-oleic sunflower seeds showed that the fat content varies in the range of 37.47−54.06% [49–51].

The ash content found in various high-oleic seeds ranged between 2.68–4.87%. The findings were lower than those reported, differences being attributed to genetic factors and geographical conditions [52,53].

Compared to the other nutrients moisture content is the most important factor for the prevention of insect infestation and diseases. Moreover, moisture content affects the physical properties of sunflower seeds. With the increase, the spatial and gravimetric also increased [39,54,55]. Moisture values found in the literature ranged between 2.5−6.32% and 3–3.2% in whole and dehulled seeds, respectively [35,56]. The higher values were found in hulls because they have a higher water absorption. Further, low moisture in kernels can be explained by the content in oil because the two liquids are immiscible [40].

When the dehulling process is applied in kernels the protein content increases up to 23.73% and ash content up to 3.31%. The findings are in accordance with those reported by other authors [35,57]. While opposite results were observed for lipid and moisture parameters. The findings indicate that the dehulling process can contribute to improving quality and reducing undesirable characteristics. However, hulls facilitate the oil extraction process. In this case, an amount of the latter should be left, but in small proportion so as not to compromise the oil quality [35].

The result obtained for kernels were similar to those obtained by other authors [58–60].

### 2.2. Classification of the Sunflower Seeds

Lipids are predominant in the kernel's cellular structure and for this reason, their mass can be considered a potential quality parameter [40]. To investigate the latter, the coefficient of correlation between the size (L, l, W, w, T, t-length, width and thickness of the whole seeds and kernels respectively), shape ($\psi$, $\psi_k$-sphericity of the whole seeds and kernels respectively; $D_e$, $D_{ek}$- equivalent diameter of the seeds and kernels) and gravimetric parameters (M, m- mass for the seeds and kernels) was calculated for a sample of 145 unsorted seeds. The results are given in Table 2. All the correlations were significant at $p < 0.05$. The L/W, L/T, W/T ratios indicate that length is more related to width and thickness, however, width is strongly related to thickness. Kernel's length, width and thickness ratio showed a low correlation with each other. The correlation coefficient for L/l, W/w and T/t ratios indicates the fact that bigger seeds when dehulled give bigger kernels. To investigate the potential correlation with the mass, there were calculated all the ratio combinations with the spatial parameters of the seeds and kernels. Mass is more related to thickness and width in the seeds and with length in the kernels. A moderate relationship between the spatial characteristics of the seeds and kernel with the mass of the kernels was found. Moreover, strong relationships were found between the seeds and the kernel's mass (r = 0.97) and between the equivalent diameter and the kernels' and seeds' mass (r = 0.84). However, the equivalent diameter is hard to obtain because it depends on the length, thickness and width. Overall, it can be concluded that there is a high linear relationship between the hulls and seeds' mass.

Considering the values obtained, the mass can be used as a parameter for the classification of sunflower seeds in three classes. The mass boundaries for each class were taken from Munder, 2017 [40]: for class I, m $\leq$ 0.045 g, for class II between 0.045−0.070 g and m > 0.070 g for class three. A proportion of 13.08% of the total sunflower seeds sample enters in the first class, a percent of 55.23% in the second and one of 21.68% in the third. Based on the lack of significance found for M/m Gupta and Das [55] choose to classify sunflower seeds based on their length. On the other hand, Santalla et al. [61] despite finding M/m the highest significant combination, choose a classification based on the length and did not justify their decision.

### 2.3. Size, Shape and Gravimetric Properties of Sunflower Seeds and Kernels

The dimensional, geometric and gravimetric properties of the four seed and kernel categories are presented in Table 3. ANOVA analysis indicates that with the increase in mass, all the shapes and spatial dimensions of the sunflower seeds and kernels have expanded significantly ($p < 0.05\%$) giving longer, wider, thicker, rounder and heavier seeds and kernels. The significant increase in bulk density and decrease in porosity with classes is correlated with the sphericity because rounder objects tend to occupy more equally the space within a given volume. As the thickness was lower than the width in both kernels and seeds, they can be described as having compressed oval bodies [40]. Further, the dimensional properties are important for determining the seed processing machines' aperture [62].

**Table 2.** Correlation between the dimensional and gravimetric properties of sunflower seeds and kernels, $n = 145$.

| Parameters | Ratio Value | Correlation Coefficient |
|---|---|---|
| L/W | 2.04 | 0.693 * |
| L/T | 3.34 | 0.567 * |
| W/T | 1.64 | 0.828 * |
| l/w | 2.17 | 0.431 * |
| l/t | 3.65 | 0.828 * |
| w/t | 1.68 | 0.404 * |
| L/M | 182.65 | 0.625 * |
| W/M | 89.69 | 0.805 * |
| T/M | 54.67 | 0.793 * |
| $D_e$/M | 96.24 | 0.841 * |
| $\psi$/M | 8.67 | 0.615 * |
| l/m | 183.77 | 0.599 * |
| t/m | 50.43 | 0.511 * |
| w/m | 84.81 | 0.626 * |
| $D_{ek}$/m | 9.21 | 0.723 * |
| $\Psi_k$/m | 1.07 | 0.259 * |
| L/m | 23.91 | 0.617 * |
| W/m | 11.74 | 0.809 * |
| T/m | 7.16 | 0.791 * |
| $D_e$/m | 12.60 | 0.840 * |
| $\psi$/m | 1.14 | 0.623 * |
| M/m | 1.31 | 0.965 * |
| L/w | 2.82 | 0.538 * |
| L/l | 1.30 | 0.453 * |
| W/w | 1.39 | 0.679 * |
| T/t | 1.42 | 0.603 * |

The symbol * indicates significance at $p < 0.01$; L—length of the whole seeds; W—width of the whole seeds; T—thickness of the whole seeds; M—mass of the whole seeds; l, w, t, m—length, width, thickness and mass of the kernels; $D_e$—equivalent diameter of the sunflower seeds; $D_{ek\text{-}}$—equivalent diameter of sunflower kernels; $\psi$, $\psi_k$—sphericity of the sunflower seeds and kernels, respectively.

The bulk density ($p_b$) through the three classes of seeds varied from 395.23 Kg/m$^3$ to 425.47 Kg/m$^3$, while kernel's density ($p_{bk}$) varied from 414.81 Kg/m$^3$ to 598.08 Kg/m$^3$. Kernel's values were higher than those of seeds, due probably to the hulls which are bulkier and provoked lower values for the mass per unit volume occupied by the seeds [55]. This characteristic depends on the distribution of seeds after shaking and the shape of the single particles, The more compacted and shaken the seeds are, the higher the values that are found for bulk density [63].

True density values for kernels ($p_{tk}$), namely 1068.60 Kg/m$^3$ to 1079.69 Kg/m$^3$, were higher than those found for seeds ($p_t$), namely 708.07 Kg/m$^3$ to 650.33 Kg/m$^3$. The finding indicated that seeds will float in water while kernels will sink [39]. Furthermore, according to this information, the separation of the hulls can be carried out by blowing air instead of floating in water [64]. The decrease in overall true density with classes may be due to the decrease in water absorption caused by the oil molecules and the increase in proteins [65]. Porosity decreased through classes from 61.58% to 44.03% and from 43.18% to 34.58% for kernels and seeds respectively. The porosity is important during the drying process because indicates the resistance of the seeds to airflow [62]. Kernels' sphericity (0.50) was lower than those of the seeds (0.53), making the seeds closer to the shape of a sphere than the kernels. However, the $\psi$ found was relatively low indicating the difficulty of the seeds to rotate easily during handling [66]. Moreover, sphericity near the value of 1 shows a higher tendency to rotate about any of the three major axes. This information is important in designing seed hoppers [62].

**Table 3.** Dimensional and gravimetric parameters for the unsorted seeds and kernels and their three categories, $n = 500$.

| Type | Parameters | Unsorted | Classification Based on Mass | | |
|---|---|---|---|---|---|
| | | | Class I | Class II | Class III |
| | | Shape and spatial dimension | | | |
| Seed | $L$, mm | $11.16 \pm 0.03$ [b] | $10.27 \pm 0.74$ [d] | $10.83 \pm 0.71$ [c] | $12.20 \pm 0.53$ [a] |
| | $W$, mm | $5.48 \pm 0.02$ [b] | $4.13 \pm 0.10$ [d] | $5.25 \pm 0.29$ [c] | $6.53 \pm 0.29$ [a] |
| | $T$, mm | $3.34 \pm 0.01$ [b] | $2.42 \pm 0.26$ [d] | $3.21 \pm 0.24$ [c] | $4.01 \pm 0.25$ [a] |
| | $D_e$, mm | $5.88 \pm 0.07$ [b] | $4.67 \pm 0.18$ [d] | $5.66 \pm 0.19$ [c] | $6.83 \pm 0.19$ [a] |
| | $\Psi$, - | $0.53 \pm 0.01$ [b] | $0.46 \pm 0.03$ [c] | $0.52 \pm 0.03$ [b] | $0.56 \pm 0.02$ [a] |
| | $V$, mm³ | $113.88 \pm 5.12$ [b] | $46.83 \pm 3.53$ [d] | $95.60 \pm 5.26$ [c] | $178.80 \pm 6.94$ [a] |
| | $S$, mm² | $110.15 \pm 2.54$ [b] | $68.59 \pm 5.16$ [d] | $100.84 \pm 6.11$ [c] | $146.47 \pm 8.27$ [a] |
| | $A_p$, mm² | $48.48 \pm 0.89$ [b] | $33.33 \pm 2.56$ [d] | $44.65 \pm 3.71$ [c] | $62.56 \pm 4.49$ [a] |
| Kernel | $L$, mm | $8.58 \pm 0.63$ [b] | $7.52 \pm 0.48$ [c] | $8.55 \pm 0.51$ [b] | $9.15 \pm 0.41$ [a] |
| | $W$, mm | $3.96 \pm 0.25$ [b] | $3.47 \pm 0.21$ [c] | $3.8 \pm 0.28$ [b] | $4.44 \pm 0.28$ [a] |
| | $T$, mm | $2.35 \pm 0.16$ [b] | $2.05 \pm 0.18$ [c] | $2.22 \pm 0.17$ [b] | $2.69 \pm 0.19$ [a] |
| | $D_{ek}$, mm | $4.30 \pm 0.13$ [b] | $3.8 \pm 0.15$ [d] | $4.15 \pm 0.11$ [c] | $4.78 \pm 0.12$ [a] |
| | $\Psi_k$, - | $0.50 \pm 0.03$ [b] | $0.50 \pm 0.03$ [b] | $0.49 \pm 0.03$ [b] | $0.52 \pm 0.02$ [a] |
| | $V_k$, mm³ | $41.05 \pm 3.56$ [b] | $26.99 \pm 1.15$ [d] | $34.99 \pm 3.31$ [c] | $57.02 \pm 4.94$ [a] |
| | $S_k$, mm² | $58.39 \pm 3.61$ [b] | $44.37 \pm 3.8$ [d] | $54.20 \pm 2.86$ [c] | $71.66 \pm 3.57$ [a] |
| | $A_{pk}$, mm² | $26.76 \pm 2.35$ [b] | $20.49 \pm 1.51$ [c] | $25.50 \pm 2.18$ [b] | $31.88 \pm 2.60$ [a] |
| | | Gravimetric properties | | | |
| Seed | $M$, g | $0.0611 \pm 0.002$ [b] | $0.0395 \pm 0.001$ [c] | $0.0569 \pm 0.003$ [b] | $0.07856 \pm 0.007$ [a] |
| | $p_b$, Kg/m³ | $404.54 \pm 2.76$ [b] | $395.23 \pm 2.53$ [c] | $415.08 \pm 2.49$ [b] | $425.47 \pm 3.13$ [a] |
| | $p_t$, Kg/m³ | $704.81 \pm 1.15$ [a] | $708.07 \pm 4.63$ [a] | $691.22 \pm 1.3$ [b] | $650.33 \pm 2.25$ [c] |
| | $\varphi$, Kg/m³ | $42.60 \pm 0.77$ [b] | $44.18 \pm 0.95$ [a] | $39.95 \pm 0.93$ [c] | $34.58 \pm 0.18$ [d] |
| Kernel | $m$, g | $0.0467 \pm 0.003$ [b] | $0.0292 \pm 0.002$ [c] | $0.0403 \pm 0.003$ [b] | $0.0591 \pm 0.004$ [a] |
| | $p_b$, Kg/m³ | $525.29 \pm 4.03$ [b] | $414.81 \pm 5.29$ [d] | $484.00 \pm 3.05$ [c] | $598.08 \pm 4.43$ [a] |
| | $p_t$, Kg/m³ | $1072.13 \pm 0.75$ [b] | $1079.69 \pm 0.45$ [a] | $1074.41 \pm 1.21$ [c] | $1068.60 \pm 0.73$ [d] |
| | $\varphi$, Kg/m³ | $51.02 \pm 0.03$ [c] | $61.58 \pm 0.02$ [a] | $54.95 \pm 0.05$ [b] | $44.03 \pm 0.04$ [d] |

$L$, $W$, $T$, $M$, $D_e$, $\psi$, $V$, $S$, $A_p$—length, width, thickness, mass, equivalent diameter, sphericity, volume, surface area and projected area of sunflower seeds; $l$, $w$, $t$, $m$, $D_{ek}$, $\psi_k$, $V_k$, $S_k$, $A_{pk}$—length, width, thickness, mass, equivalent diameter, sphericity, volume, surface area and projected area of sunflower kernels; $p_b$—bulk density; $p_t$—true density; $\varphi$—porosity. Different superscript letters indicated difference at $p < 0.05\%$.

Values found by others authors for the seed's bulk density, true density and porosity ranged between 267.03–710 Kg/m³, 444.39–902 Kg/m³ and 31.3–54.93% respectively. While those for kernels ranged between 535–582.50 Kg/m³, 1015–1250 Kg/m³ and 45.4–51.19% [37,39–41,54,55,61,67,68].

All seeds categories are longer, wider and thinner compared with the Modern sunflower variety [55]. In comparison with the sunflower hybrid F1 from cultivar PR65H22, the seeds presented similar lengths, but they were thicker and wider [40]. Moreover, the findings showed smaller, wider and thicker seeds than Trisum 568 sunflower genotype [61]. Compared with the two sunflower hybrids (ACA 884 and Paraiso 20) selected by de Figueiredo [41], the unsorted seeds were longer, wider and thicker. The opposite was observed when unsorted seeds were compared with the PSH-996, Shamshiri and the P64H41 varieties [37,39,54]. From the six varieties studied by Cetin,2020, [49] Transol and Colombi showed higher dimensional values, while Tunca presented a similar dimension, except for the length, which was longer.

For the unsorted category, the equivalent diameter, sphericity, volume and surface area values were higher than those reported for the PSH-996 variety. However, they were lower than those reported for Shamshiri, P64H41, LG5582, Transol, 63MM54, P64LC53, Colombi varieties. The three categories of sunflower seeds presented slightly higher sphericity, diameter and volume values than those reported for the PR65H22 variety.

*2.4. Sunflower Oilcakes Characterization*

The physical, chemical and functional properties of the two types of cold-pressed sunflower oilcakes are reported in Table 4. The analyzed sunflower oilcakes have different shapes, namely pellets (SFOC/PE) and cake (SFOC/C). The nutritive composition of the sunflower oilcakes can differ considerably depending on the quality of seeds, extraction technique and storage parameters. All the findings fell in the range reported by other authors [69–71]. No significant difference ($p < 0.05$) in the moisture and protein content was found between the two samples. SFOC/C presented significantly higher ash and crude fiber values, but lower fat content than SFOC/PE.

The moisture content is an important factor to maintain oilcake stability for long periods of time [72]. A level below 12% is considered safe for storage because it prevents the rapid growth of mold [62]. The values obtained were 8.75% for the meal pellets and 8.93% for the meal cake. The values were relatively similar to those reported for soybean, rapeseed, sesame and flaxseed. Much lower values were found for hemp seed and pumpkin [69,72–74].

Oilcakes should be admitted for human consumption when there is an equilibrated proteins and lipids ratio, optimal values for the human body should be 20–25% and 3–5%, respectively [75]. In our case, the fat amounts are too high so direct consumption is impossible. Thus, sunflower oilcakes are destined for the extraction of bioactive compounds.

The total dietary fiber content in the two sunflower press-cakes was high (31.88% for SFOC/PE and 12.64% FOR SFOC/C). The findings met consumers' demands for fiber-rich food. In addition, fibers have numerous beneficial effects (increase laxation and decrease blood pressure, cholesterol level, reabsorption of bile acids and starch digestion) [76–79].

Water retention capacity (WRC) offers information about the degradation of the molecular components by measuring the amounts of solid components released from proteins and other molecules. The WRC for the two sunflower press-cakes was 4.67 g/g for the one in pellet shape and 5.51 g/g for the cake, the difference between the two was not significant ($p < 0.05$) and was probably due to the moisture and protein contents. The same results were found by other authors and ranged from 2.10 g/g to 4.48 g/g [80,81].

The capacity to absorb oil and water is determined by the non-polar and polar amino acids composition, respectively. The oil holding capacity of SFOC/PE was slightly higher than those of SFOC/C, but the difference was not significantly different (confidence level of 95%). A small amount of lipids and moisture increases protein solubility and thus the absorption capacity of the oilcakes [80]. In other studies, values for OHC that ranged from 0.71 g/g to 2.2 g/g [80–85] were found.

Water holding capacity (WHC) is measured by the amount of water absorbed by the molecules. The parameter is important for determining the storage conditions. The difference found between the two oilcakes was significant ($p < 0.05$) [79]. SFOC/C presented higher moisture content that provoked a reduction of the degradation of the molecule and hence the reduction of the parameter. Another possible explanation for the highest values found in SFOC/PE refers to the high content of dietary fibers [85]. WHC values found in our study were similar to those reported by other authors, which ranged from 0.71 g/g to 3.27 g/g [80–85].

The bulk density (BD) of the two oilcakes was not significantly different from each other. The index decreased with moisture and increased when lipids content decreased. BD is an important property in the packaging and handling processes in the food industry. It is a measure of flour heaviness and depends on the attractive intermolecular forces, particle size and number of positions in connection [72,86]. Other values found in literature ranged from 0.592 g/mL to 0.741 g/mL [80,87]

**Table 4.** Chemical, functional and color proprieties of the two sunflower oilcakes.

| Parameters | SFOC/PE | SFOC/C |
|---|---|---|
| PHYSICO-CHEMICAL PROPERTIES | | |
| Moisture (%) | 8.75 ± 0.10 [a] | 8.93 ± 0.11 [a] |
| Dry matter (%) | 91.25 ± 0.10 [a] | 91.07 ± 0.11 [a] |
| Proteins (%) | 20.15 ± 1.57 [a] | 21.60 ± 1.87 [a] |
| Fat (%) | 15.77 ± 0.45 [a] | 14.16 ± 0.04 [b] |
| Ash (%) | 4.56 ± 0.11 [b] | 6.15 ± 0.04 [a] |
| Crude fiber (%) | 31.88 ± 0.79 [a] | 12.64 ± 0.05 [b] |
| Carbohydrates (%) | 18.89 ± 0.23 [a] | 36.52 ± 1.11 [b] |
| FUNCTIONAL PROPERTIES | | |
| Bulk density (g/mL) | 0.4196 ± 0.002 [a] | 0.4204 ± 0.001 [a] |
| WHC (g/g) | 2.58 ± 0.11 [a] | 2.33 ± 0.07 [b] |
| OHC (g/g) | 1.34 ± 0.13 [a] | 1.18 ± 0.08 [a] |
| WRC (g/g) | 4.67 ± 0.04 [a] | 5.51 ± 0.06 [a] |
| SC (%) | 3.56 ± 0.06 [a] | 3.19 ± 0.17 [a] |
| EC (%) | 32.17 ± 1.15 [a] | 30.62 ± 2.14 [a] |
| ES (%) | 29.87 ± 1.24 [a] | 27.92 ± 0.57 [a] |
| COLOUR PROPERTIES | | |
| L* | 42.23 ± 0.01 [b] | 46.29 ± 0.01 [a] |
| a* | 1.17 ± 0.01 [b] | 1.57 ± 0.01 [a] |
| b* | 6.11 ± 0.01 [b] | 8.90 ± 0.01 [a] |

SFOC/PE—sunflower oilcakes in pellets form; SFOC/C—sunflower oilcake in the form of cake; WHC—water holding capacity; OHC—oil holding capacity; WRC—water retention capacity; SC—swelling capacity, EC—emulsion capacity; ES—emulsion stability; L*—lightness; a*—redness; b*—yellowness. For difference assessment was performed Student t-test. Values followed by [a], [b] are statistically different at 95% confidence level.

Emulsion capacity (EC) is the property of mixing two immiscible liquids (water and oil). Emulsion stability (ES) measures the amount of water released from the emulsion over time. The parameters are closely related to protein surface hydrophobicity that allows a better molecular anchorage of the oil–water interface and thus more stable emulsions [72]. SFOC/C presented a lower value for EC and ES that might be due to its lower amount of hydrophobic amino acid. In the literature, higher values for EC (49.09–53.2%) and ES (48.23–50.45%) [88,89] were found.

In terms of color, the SFOC/C was significantly lighter (L* = 46.29) than SFOC/PE, with higher redness (a* = 1.57) and yellowness (b* = 8.90) values. Compared to the sunflower oil analyzed by Grasso et al. [85] our meals are lighter, less red and more yellow. The color of the products obtained in the previous study was influenced when the content of 18% sunflower oilcake was added, the color becomes browner.

The values of all the nutritional parameters analyzed for the whole seeds and press cakes presented differences with each other ($p$ = 95%). Sunflower seeds had a higher caloric value than the oilcakes due to their higher lipid content. In the meals, the proportion in seeds reached 65.42% and decreased to 14.16–15.77%, once oil extraction by cold pressing was realized. Cold extraction is the most preferable method to obtain high-quality virgin oil. However, an important oil fraction remains in the press-cakes and increases their nutritional values because it provides the whole health benefits to these by-products [90]. The remaining parameters (ash, protein, carbohydrates) increased in the oilcakes as a result of oil removal [90].

Fat content in press-cakes was still high (15.77% in pellets and 14.16% in cake) and needed a further re-extraction. This can be realized with solvents or hot temperature pressing. The first allows to obtain higher oil yields but can compromise oilcake quality, while the second can lead to the liberation of aroma compounds and dark colors [90].

In conclusion, the cold oil pressing by-products are characterized by high nutritional values and good functional parameters. Press-cakes can be used as a food ingredient or for the extraction of bioactive compounds that can be incorporated in new foodstuffs because they are nutritional, social and economically advantageous [91].

Oilcakes rich in proteins and lipids are suitable for feeding omnivores and fish while being rich in fibers for ruminants. Studies showed that sunflower oilcakes improved the carcass yield of fish and pigs [92,93].

A possible valorization of sunflower seeds involves the realization of new food products such as tablets that can be used as supplements [70] Other products obtained with sunflower oilcakes were biscuits (higher protein, phenols and antioxidants compounds) [85], cookies (addition of 10% results in better proteins digestibility and water absorption) and muffins (products with low carbohydrate content) [14].

Furthermore, proteins extracted from sunflower oilcakes can be used for the production of films with good adhesive and barrier properties and low elongation, deformation and elasticity [9].

### 2.5. Comparison of the Mineral Composition of the Sunflower Seeds, Oil and Oilcake

Minerals are inorganic nutrients essential for the maintenance of life physicochemical processes [94]. They can be classified into macro-elements (potassium, phosphorus, calcium, sodium and chloride) and micro-elements (iron, copper, zinc, molybdenum, chromium, manganese, copper and selenium). The required amounts in diets for macro-elements must be greater than 100 mg/dL and less for micro-elements [95].

Sunflower seeds, oil and meal are known to be a source of several minerals [10,11,19,34]. The mineral composition is shown in Table 5. A total of 18 elements were found in seeds (Mg < Se < Ce < Ca < Tl < Zn < Mn < Cr < Cu < Ni < Be < Co < Ti < Fe < Li < Mo < Cd) and SFOC/C (Se < Ce < Ca < Tl < Zn < Cu < Mn < Cr < Sr < Be < Ni < Co < Fe < Mo < Li < Cd). In SFOC/PE were found 20 elements as follows: magnesium (4.76 g/Kg), selenium (1.99 g/Kg), cesium (1.02 g/Kg), calcium (1163.32 mg/Kg), thallium (587.97 mg/Kg), zinc (94.78 mg/Kg), strontium (72.97 mg/Kg), copper (61.15 mg/Kg), manganese (57.62 mg/Kg), chromium (52.79 mg/Kg), beryllium (32.96 mg/Kg), nickel (29.38 mg/Kg), titan (16,10 mg/Kg), cobalt (7.63 mg/Kg), iron II and III (5.26 mg/Kg, 3.35 mg/Kg), molybdenum (0.43 mg/Kg), lithium (0.34 mg/Kg), cadmium (0.23 mg/Kg) and antimony (0.02 mg/Kg). They were found only 14 elements in SFO of which thallium, cesium, magnesium and selenium presented high values, the others (Mo < Mn < Be < Cr < Cu < Li < Ni < Zn < Ti < Fe) presented proportions below 1%.

After oil extraction, most mineral composition of oilcakes increased, while Fe, Co and Li decreased. A low percentage of the elements goes into the oil. A comparison between the whole seeds, press-cakes and oil regarding the mineral composition reveals that press-cakes are richer and they are a valuable ingredient for new food products development. The results obtained were in accordance with those provided by other authors [53,96].

Calcium, cobalt, strontium, cadmium and antimony were the only elements that were not found in the oil, despite being present in sunflower seeds and oil. The elements presented only in sunflower oilcakes were strontium and antimony. The difference in elements composition was significant (95% confidence level) for all the samples studied.

### 2.6. Fatty Acids Profile of Sunflower Seeds, Oil and Oilcakes

Fatty acids (FA) composition of the seeds, cakes and oil are shown in Table 6. They were quantified using a gas chromatograph coupled with mass spectrometry. The difference between the seeds, oil and oilcakes was significant ($p < 0.5$%). A total of 14 fatty acids were determined, of which five were saturated (SFA), five monounsaturated (MUFA) and four polyunsaturated (PUFA). The total concentration of FA in seeds and oil were 440.62 µg/mL and 441.31 µg/mL, respectively, while in press-cakes were between 1016.52 µg/mL and 3083.38 µg/mL. The sunflower seeds and oil were rich in PUFA (51.41% and 64.81% respectively) and MUFA (41.69% and 20.58%) and poor in saturated FA (6.90% and 14.61%). The two oilcakes were a rich source of unsaturated FA, namely 29.46% and 57.96% monounsaturated, also 66.13% and 34.16% polyunsaturated. The most abundant FA were linoleic, pentadecanoic, stearic and oleic. Low levels were detected for palmitic, palmitoleic, heptadecenoic, linolelaidic, linolenic, eicosenoic, arachidonic and trisanoic fatty acids.

**Table 5.** Mineral composition of the sunflower seeds, oil and press-cakes.

| Parameters | SFS mg/Kg | SFOC/PE mg/Kg | SFOC/C mg/Kg | SFO mg/Kg |
|---|---|---|---|---|
| Li | 1.80 ± 0.01 [a] | 0.34 ± 0.01 [c] | 1.40 ± 0.0 [b] | 0.20 ± 0.01 [c] |
| Be | 20.89 ± 0.14 [c] | 32.96 ± 0.55 [a] | 31.48 ± 0.06 [b] | 0.72 ± 0.03 [d] |
| Mg | 3.89 ± 0.24[1] [b] | 4.76 ± 0.13[1] [a] | - | 3.44 ± 0.15 [c] |
| Ca | 573.02 ± 4.73 [c] | 1163.32 ± 10.01 [b] | 1522.08 ± 5.5 [a] | - |
| Ti | 7.04 ± 0.25 [c] | 16.10 ± 0.26 [b] | 18.38 ± 6.22 [a] | 0.03 ± 0.0 [d] |
| Cr | 35.70 ± 0.1 [c] | 52.79 ± 0.38 [b] | 58.24 ± 1.81 [a] | 0.50 ± 0.01 [d] |
| Mn | 36.91 ± 0.38 [c] | 57.62 ± 0.21 [b] | 65.73 ± 3.46 [a] | 0.78 ± 0.03 [d] |
| Fe (II) | 6.66 ± 0.13 [a] | 5.26 ± 0.20 [b] | 4.71 ± 2.71 [c] | 0.01 ± 0. [d] |
| Fe (III) | 6.40 ± 0.44 [a] | 3.35 ± 0.17 [b] | 2.51 ± 0.03 [c] | - |
| Co | 11.46 ± 0.69 [a] | 7.63 ± 0.41 [b] | 5.59 ± 0.35 [c] | - |
| Ni | 21.29 ± 1.30 [c] | 29.38 ± 1.27 [b] | 30.63 ± 1.40 [a] | 0.19 ± 0.01 [d] |
| Cu | 32.57 ± 1.79 [c] | 61.15 ± 4.12 [b] | 71.25 ± 3.36 [a] | 0.21 ± 0.01 [d] |
| Zn | 57.83 ± 2.54 [c] | 94.78 ± 2.28 [b] | 90.11 ± 4.36 [a] | 0.09 ± 0.0 [d] |
| As | - | - | - | - |
| Se | 1.22 ± 0.02[1] [c] | 1.99 ± 0.07[1] [b] | 3.18 ± 0.18 [1] [a] | 1.17 ± 0.0 [d] |
| Sr | - | 72.97 ± 2.42 [a] | 35.42 ± 1.99 [b] | - |
| Mo | 0.34 ± 0.0 [c] | 0.43 ± 0.01 [c] | 1.62 ± 0.55 [a] | 0.90 ± 0.01 [b] |
| Cd | 0.16 ± 0.00 [c] | 0.23 ± 0.01 [b] | 0.26 ± 0.0 [a] | - |
| Sb | - | 0.02 ± 0.0 [a] | - | - |
| Ce | 0.33 ± 0.01 [1 d] | 1.02 ± 0.01 [1 b] | 1.85 ± 0.05 [1 a] | 6.39 ± 0.35 [c] |
| Tl | 523.84 ± 9.11 [b] | 587.97 ± 17.80 [a] | 417.12 ± 7.31 [d] | 175.69 ± 4.56 [c] |

The superscript number [1] indicates that the results are expressed as g/Kg. The analysis was performed in duplicate. Results are presented as values ± standard deviation. SFS—sunflower seeds, SFOC/PE—pellets sunflower oilcakes, SFOC/C—sunflower oilcakes in the form of cake, SFO—sunflower oil, Li—lithium, Be—beryllium, Mg—magnesium, Ca—calcium, Ti—titan, Cr—chromium, Mn—manganese, Fe—iron, Co—cobalt, Ni—nickel, Cu—copper, Zn—zinc, As—arsenic, Se—selenium, Sr—strontium, Mo—molybdenum, Cd—cadmium, Sb—antimony, Ce—cesium, Tl—thallium. Lowercase letters ([a],[b],[c] and [d]) refer to the comparison of the same element between the different samples; results followed by superscript letters are significantly different ($p < 0.05\%$) according to Turkey's post hoc test.

Palmitic and stearic acid values varied between 2.18% and 5.52%, also between 1.10% and 10.45%, respectively. Values below 1% were found for myristic, linolelaidic, linolenic, eicosenoic, trisanoic and arachidonic fatty acids. The major FA in seeds, cakes and oil were linoleic (50.32%, 32.81–65.88% and 64.35%, respectively) and oleic (14.10%, 10.34–19.32% and 19.81%, respectively). These results make the three products (seeds, oil and oilcakes) important dietary sources of linoleic and oleic fatty acids.

Linolenic and linoleic acids are polyunsaturated essential fatty acids. They can not be synthesized by the organism and play an important role in the maintenance of healthy triglyceride and cholesterol levels. Sunflower linoleic/linolenic FA ratios were high due to the low levels of linolenic acid. Regarding the SFA, the content was relatively low ($\leq 14.61\%$). The consumption of sunflower oil, extracted or naturally presented in seeds or press-cakes, can help to increase the level of linoleic acid in the human body.

Sunflower seeds in the literature presented 9.63–10.11% saturated fatty acids, 20.73–25.77% monounsaturated fatty acids and 65.59–69.64% polyunsaturated fatty acids [59]. In sunflower oil, myristic (<0.2%), palmitic (5–7.6%), palmitoleic (<0.3%), oleic (14.1–39.4%), linoleic (48.3–74%), linolenic (<0.3%) and eicosenoic (<0.5%) were found. All the results are in accordance with those obtained in our study [59].

**Table 6.** Fatty acids composition of sunflower seeds, oil and oilcake.

| Fatty Acid [1] | Type | SFS [2] µg/mL | Relative Level [3] % | SFOC/PE µg/mL | Relative Level % | SFOC/C µg/mL | Relative Level % | SFO µg/mL | Relative Level % |
|---|---|---|---|---|---|---|---|---|---|
| C14:0 | SFA | 2.25 ± 0.20 [c] | 0.43 ± 0.02 [A] | 5.05 ± 0.02 [a] | 0.16 ± 0.04 [C] | 3.78 ± 0.02 [b] | 0.31 ± 0.00 [B] | - | - |
| C15:1 | MUFA | 47.98 ± 0.20 [b] | 27.60 ± 0.50 [B] | - | - | 148.28 ± 0.36 [a] | 36.82 ± 0.09 [A] | - | - |
| C16:0 | SFA | 38.46 ± 0.11 [d] | 2.44 ± 0.12 [C] | 210.89 ± 1.35 [a] | 2.18 ± 0.12 [D] | 201.91 ± 0.54 [b] | 5.52 ± 0.32 [A] | 41.07 ± 0.04 [c] | 3.41 ± 0.12 [B] |
| C16:1 | MUFA | - | - | 0.96 ± 0.00 [c] | 0.03 ± 0.00 [C] | 123.01 ± 0.95 [a] | 10.08 ± 0.53 [A] | 3.08 ± 0.07 [b] | 0.77 ± 0.04 [B] |
| C17:1 | MUFA | - | - | 1.03 ± 0.04 [b] | 0.06 ± 0.00 [B] | 4.38 ± 0.02 [a] | 0.72 ± 0.04 [A] | - | - |
| C18:0 | SFA | 26.47 ±0.07 [c] | 2.51 ± 0.06 [B] | 114.30 ± 0.98 [a] | 1.77 ± 0.15 [C] | 26.91 ± 0.18 [c] | 1.10 ± 0.09 [D] | 83.77 ± 0.54 [b] | 10.45 ± 0.34 [A] |
| C18:1 (w-9) | MUFA | 49.47 ± 0.20 [d] | 14.10 ± 0.32 [C] | 629.87 ± 0.35 [a] | 29.32 ± 0.95 [A] | 84.06 ± 0.54 [b] | 10.34 ± 0.45 [D] | 52.97 ± 0.32 [c] | 19.81 ± 0.54 [B] |
| C18:2 (all-trans 9,12) (w-6 t) | PUFA | 3.98 ± 0.01 [b] | 0.76 ± 0.03 [A] | 2.90 ± 0.07 [c] | 0.09 ± 0.04 [C] | 8.17 ± 0.03 [a] | 0.67 ± 0.04 [B] | - | - |
| C18:2 (all-cis 9,12) (w-6) | PUFA | 262.34 ± 0.11 [c] | 50.32 ± 1.49 [C] | 2102.26 ± 5.55 [a] | 65.88 ± 1.55 [A] | 396.30 ± 0.17 [b] | 32.81 ± 1.85 [D] | 255.59 ± 1.32 [d] | 64.35 ± 0.14 [B] |
| C18:3 (w-3) | PUFA | 1.70 ± 0.02 [c] | 0.33 ± 0.04 [B] | 5.35 ± 0.15 [a] | 0.17 ± 0.00 [C] | 2.17 ± 0.12 [b] | 0.18 ± 0.04 [C] | 1.83 ± 0.42 [c] | 0.46 ± 0.02 [A] |
| C20:1 (w-9) | MUFA | - | - | 1.25 ±0.04 [a] | 0.04 ± 0.00 [A] | - | - | - | - |
| C20:4 (w-6) | PUFA | - | - | - | - | 6.07 ± 0.22 [a] | 0.50 ± 0.03 [A] | - | - |
| C21:0 | SFA | 7.99 ± 0.07 [b] | 1.53 ± 0.00 [A] | 6.68 ± 0.98 [c] | 0.21 ± 0.07 [D] | 11.50 ± 0.54 [a] | 0.95 ± 0.04 [B] | 3.00 ± 0.22 [d] | 0.75 ± 0.04 [C] |
| C23:0 | SFA | - | - | 2.86 ± 0.07 [a] | 0.09 ± 0.04 [A] | - | - | - | - |
| C18:2 w-6/C18:3 w-3 | | 152.49 ± 0.98 [C] | | 387.53 ± 4.55 [A] | | 182.28 ± 1.95 [B] | | 139.89 ± 0.00 [D] | |
| C18:1 w-9/C18:2 w-6 | | 0.28 ± 0.01 [C] | | 0.45 ± 0.01 [A] | | 0.32 ± 0.001 [B] | | 0.31 ± 0.00 [B] | |
| ΣSFAs (%) | | 6.90 ± 0.04 [C] | | 4.41 ± 0.09 [D] | | 7.88 ± 0.07 [B] | | 14.61 ± 0.04 [A] | |
| ΣUFAs (%) | | 93.1 ± 0.54 [B] | | 95.59 ± 0.98 [A] | | 92.12 ± 1.55 [C] | | 85.39 ± 1.49 [D] | |
| ΣMUFAs (%) | | 41.69 ± 1.54 [B] | | 29.46 ± 0.54 [C] | | 57.96 ± 0.32 [A] | | 20.58 ± 0.17 [D] | |
| ΣPUFAs (%) | | 51.41 ± 1.32 [C] | | 66.13 ± 0.25 [A] | | 34.16 ± 0.15 [D] | | 64.81 ± 0.35 [B] | |
| ΣSFAs/ΣUFAs | | 0.07 ± 0.00 [C] | | 0.05 ± 0.00 [C] | | 0.09 ± 0.00 [B] | | 0.17 ± 0.00 [A] | |

[1] C14:0, Myristic; C15:1, Pentadecenoic; C16:0, Palmitic; C16:1, Palmitoleic; C17:1, Heptadecenoic; C18:0, Stearic; C18:1, Oleic; C18:2 (all trans 9.12), Linolelaidic; C18:2 (all-cis 9.12), Linoleic; C18:3, Linolenic; C20:1, Eicosenoic; C20:4, Arachidonic; C21:0, Heneicosanoic acid; C23:0 Trisanoic acids. SFAs, saturated: UFAs, unsaturated; MUFAs, monosaturated; PUFAs, polyunsaturated fatty acids. [2] Values are presented as mean± standard deviation. When followed by different superscript letters ([a], [b], [c], [d]) they are statistically different at 95% confidence level. [3] Uppercase superscript refers to relative level %. Values with different letters ([A], [B], [C], [D]) are statistically different at $p < 0.05\%$

Values for sunflower meals found in the literature ranged between 11.3%–67.82% for SFA, between 20.6%–25.90% for MUFA and between 3.81–68.2% for PUFA [13]. Regarding the C18:2 w-6/C18:3 w-3 ratio found in our study was in accordance with the values found in the literature (3.86–37.79) [97]. The fatty acids profile in the literature include myristic (0.30%–9.63%), palmitic (12.05–29.1%), stearic (12.2%), linolelaidic (0.04%), linoleic (1.93–57.82%), linolenic (0.39–1.53%), eicosenoic (0.04%), arachidonic (0.02%) and trisanoic (0.05%) fatty acids [98]. Values were in accordance with the results obtained in our study.

In conclusion, sunflower oilcakes can be used for the development of new food products due to their advantageous FA profile, where oleic acid is predominant.

### 2.7. Amino Acids Profile of Sunflower Seeds, Oil and Oilcakes

Proteins presented the highest increase in press-cakes. To evaluate their quality the amino acids (AA) profile must be determined (Table 7).

**Table 7.** Amino acids profile of sunflower seeds, oil and oilcake.

| Parameters | SFS<br>nmol/g | SFOC/PE<br>nmol/g | SFOC/C<br>nmol/g |
|---|---|---|---|
| Alanine | 2110.4 ± 18.21 [b] | 2187.18 ± 36.93 [b] | 3073.51 ± 43.48 [a] |
| Glycine | 1810.93 ± 0.0 [a] | 2329.15 ± 0.0 [a] | 1696.04 ± 324.38 [a] |
| Valine * | - | 8987.78 ± 3.40 [a] | 905.97 ± 22.10 [b] |
| Leucine * | 383.84 ± 0.0 [a] | 164.77 ± 4.20 [b] | 486.66 ± 8.00 [a] |
| Isoleucine * | - | 1584.28 ± 14.59 [a] | 757.88 ± 2.04 [b] |
| Threonine * | - | 827.21 ± 17.53 [a] | 742.32 ± 8.33 [b] |
| Serine | - | 2124.69 ± 12.66 [a] | 1181.12 ± 26.93 [b] |
| Proline | - | 1313.13 ± 8.66 [a] | 887.04 ± 12.66 [b] |
| Asparagine | - | 1102.81 ± 10.90 [a] | 629.15 ± 3.76 [b] |
| Aspartic acid | 255.53 ± 3.58 [c] | 2949.61 ± 137.36 [a] | 2045.89 ± 19.58 [b] |
| Methionine * | - | 696.83 ± 3.92 [a] | 675.17 ± 1.90 [b] |
| Phenylalanine * | - | 768.82 ± 1.56 [a] | 757.25 ± 8.43 [b] |
| Glutamic acid | 1229.56 ± 0.0 [a] | 3402.01 ± 0.0 [b] | 2082.36 ± 39.83 [a] |
| α-aminoadipic acid | - | - | 680.19 ± 18.43 [a] |
| Hidroxylysine | - | - | 686.60 ± 33.33 [a] |
| Tyrosine | - | - | 650.22 ± 0.32 [a] |
| Tryptophan * | - | - | 1093.97 ± 14.53 [a] |
| Total, nmol | 5790.26 [c] | 28438.27 [a] | 19031.34 [b] |
| Essential AA, % | 6.63 [c] | 45.82 [a] | 28.48 [b] |
| Non essential AA, % | 93.37 [a] | 54.18 [c] | 71.52 [b] |

The symbol * indicates an essential amino acid. The analysis was performed in duplicate. Values are presented as mean ± standard deviation. Values followed by different superscript letters ([a], [b], [c], [d]) are statistically different at 95% confidence level.

The findings showed higher total amino acids content in oilcakes than seeds, 28438.27 nmolg$^{-1}$, 19031.34 nmolg$^{-1}$ and 5790.26 nmolg$^{-1}$, respectively. In descending order, the amino acids identified in seeds were alanine, glycine, glutamic acid, leucine and aspartic acid. Asparagine and glutamine were not found because they were totally converted to aspartic and glutamic acids in acidic hydrolysis conditions.

In pellets press-cake were found the following 13 AA: valine, glutamic acid, aspartic acid, glycine, alanine, serine, isoleucine, proline, asparagine, threonine, phenylalanine, methionine and leucine. On the other hand, in the cake meal were found 17 AA namely, alanine < glutamic acid < aspartic acid < glycine < serine < tryptophan < valine < proline < isoleucine < phenylalanine < threonine < hydroxylysine < α-aminoadipic acid <methionine < tyrosine < asparagine < leucine.

In cakes, essential AA represents 45.82% and 28.48% of the total AA profile, while in seeds only 6.63%. Valine and tryptophan were the major essential AA found in meals. They were followed by isoleucine, threonine and phenylalanine. All the essential AA must be obtained through an equilibrate diet because they cannot be synthesized in the human body.

Sunflower press cake in the form of pellets presented the most amino acids (28438.27 nmol/g), the difference between all the products was significant ($p$ < 95%).

For all the amino acids found in the samples were calculated the percentage relative level which is shown in Figure 1. Based on the relative percentages were calculated the total percentage of essential and non-essential AA (Table 7).

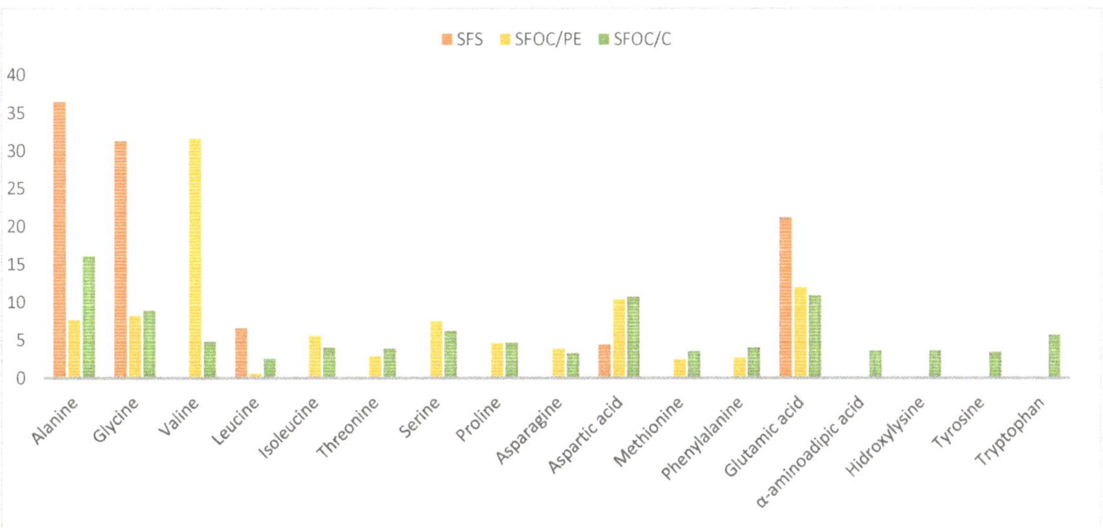



**Figure 1.** Amino acids of sunflower seeds, meals and oil.

Alanine (difference was significant), glycine (difference was not significant) valine (significant difference) and glutamic acid (significant difference) were the most predominant amino acids, while leucine was the least (significant difference). A total of twelve amino acids (valine, isoleucine, threonine, serine, proline, asparagine, methionine, phenylalanine, α-aminoadipic acid, hidroxylysine, tyrosine and tryptophan) were found in meals but not in the seeds. This is due to the presence of the hulls, which create an intercellular skeleton that prevents the action of digestive enzymes and thus reduces the amino acids present in the hull [99].

### 3. Materials and Methods

*3.1. Samples*

For this study, high oleic sunflower seeds (*Helianthus annuus L.*), oilcakes and oil were a kind gift from a local factory of Suceava (Romania), that uses traditional mechanical oil pressing technology. The seeds (SFS) (Figure 2) were cleaned manually to remove foreign materials and immature seeds. Kernels and hulls were obtained by manually dehulling.

The two oilcakes have different forms: pellets (SFOC/PE) and cake (SFOC/C) (Figure 2). They were ground and sieved (< 500 μm) with a Retsch Vibratory Sieve Shaker AS 200 basic (Retsch GmbH, Haan, Germany).

*3.2. Chemical Composition*

3.2.1. Seeds and Kernels

Moisture content was measured using a gravimetric method (ISO 665:2020) by drying a 3 g sample in the laboratory oven with air circulation ZRD-A5055 (Zhicheng Analysis Instruments, Shanghai, China) at 105 ± 2 °C for 24 h.

**Figure 2.** Sunflower seeds and press-cakes.

Ash content was estimated with AOAC method 923.03 using a calcination furnace at 550 °C for 6 h till the charred material became white [74].

Protein content was determined using the Kjeldahl method (AOAC 950.48) described by Sunil, 2015 [74] with some modifications and the conversion factor 5.88. Digestion was made as follows: 1 g of sample with 15 mL of sulfuric acid and a catalyst were placed in a Kjeldahl flask. Digestion was carried out at 420 °C for 2 h until a clear blue solution was obtained. Afterward, 50 mL of distilled water is added to the flask. The distillation was carried out with boric acid, while titration with hydrochloric acid was 0.2 N.

The lipid content of seeds, kernels and hulls was determined by ISO 659:2009 using an automatic Soxhlet extraction system with n-hexane solvent.

### 3.2.2. Oilcakes

The oilcakes were investigated for their moisture, ash, fat, protein and total dietary fiber (Megazyme total dietary fiber assay kit, Megazyme, Wicklow, Ireland) with AOAC methods with some modifications (methods 935.29, 923.03, 920.39, 950.48 and 985.29 respectively).

The color of the press-cakes was measured using a CR-400 colorimeter (Konica Minolta, Tokyo, Japan) and CIELAB scale: lightness L* (0 for black and 100 for white), a* (if negative indicate the intensity of green, if positive of red) and b*(if negative indicate the intensity of blue, if positive of yellow) [100].

The water/oil holding capacity (WHC, OHC) was measured according to the method described by Omowaye-Taiwo, 2015 [101], with slight modifications. In test tubes, 1 g of sample and 10 mL of distilled water/corn oil were added. Then they were kept at room temperature for 30 min and centrifugated at 7000 rpm for 20 min. The results were expressed as grams of water/oil absorbed per gram of sample.

Water retention capacity (WRC) was determined according to the method described by Onipide et al., 2017 [102]. The sample (2 g) was mixed with 20 mL distilled water, kept for 1 h at 25 °C in a water bath (Memmert Waterbath WNB 22, Schwabach Germany) with continuous stirring and centrifugated at 1600 rpm for 25 min. The supernatant was removed and the samples were weighed. WRC was calculated as the difference between the hydrated and dry residues.

To measure the swelling capacity (SC), 1 g of sample and 10 mL of distilled water were mixed in a centrifuge tube. After centrifugation for 20 min at 2000 rpm, the supernatant was decanted and the remained sample was dried in a hot air oven for 2 h at 130 °C and then weighted. The results were expressed as percent swelled per gram sample [103].

Bulk density (BD) of sunflower meals was analyzed by a volumetric method; 5 g of sample were placed in a 100 mL cylinder and gently tapped 20 times. The values were calculated as sample weight and volume displaced ratio (g/mL) [80].

Emulsion capacity (EC) and stability (ES) were analyzed according to Rani, 2021 [72] and Yyenagbe et al., 2017 [104] as follows: a 0.5% suspension of sample and water was mixed on a magnetic stirrer for 20 min at 500 rpm, 30 mL of this emulsion were then taken and mixed with corn oil (10 mL). The emulsion was homogenized and transferred immediately to a marked cylinder (50 mL) in order to read the height obtained. The ES was calculated by heating the cylinder for 30 min at 80 °C. After that, the final height of the emulsion was read. The formulas for calculating EC and ES are expressed in Equations (1) and (2).

$$EC\ (\%)\ =\ \frac{\text{Height of oil layer}}{\text{Height of the suspension}}\ \times\ 100 \tag{1}$$

$$ES\ (\%)\ =\ \frac{\text{Final height of the emulsion layer}}{\text{Initial height of the emulsion layer}}\ \times\ 100 \tag{2}$$

*3.3. Physical Properties of Seeds and Kernels*

3.3.1. Mass and Classification

A group of randomized whole seeds was weighted with an analytical balance PART-NER AS 220.R2 (Radwag, Bucharest, Romania) at 0.1 mg accuracy. In this way, the percentage of hull and kernel was determined. Moreover, based on mass, the kernels and seeds were classified into three categories [40].3.3.2. Dimensional Parameters

Groups of 100 seeds were randomly measured to determine their dimensions. Length (L), width (W) and thickness (T) were analyzed using a digital caliper VOREL 15240 (Toya, Wrocław, Poland) at 0.003 mm accuracy [40]. In the same way, the size and shape (l, w and t) of kernels were determined (Figure 3).

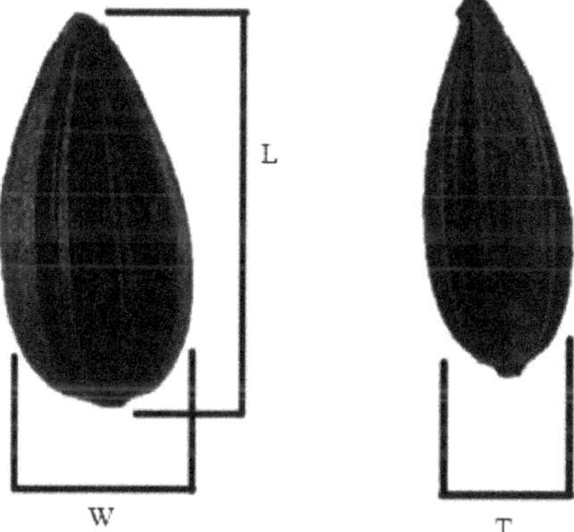


**Figure 3.** Spatial dimension of sunflower seeds; L—length, W—width, T—thickness.

3.3.2. Geometric Parameters

The equivalent diameter ($D_e$), sphericity ($\psi$), seed surface area (S), projected area ($A_p$) and volume (V) were determined by comparison to a sphere using the following relationships [37,39,40,49]:

$$De\ (\text{mm})\ =\ \sqrt[3]{L\ \times\ W\ \times\ T} \tag{3}$$

$$\psi\ =\ D_e/L \tag{4}$$

$$S\ \left(\text{mm}^2\right)\ =\ \Pi\ \times\ D_e{}^2 \tag{5}$$

$$A_p\ \left(\text{mm}^2\right)\ =\ {}^3\!/_4\ \times\ L\ \times\ W \tag{6}$$

$$V\left(\text{mm}^3\right)\ =\ W\ \times\ L\ \times\ T\ \times\ \varphi \tag{7}$$

### 3.3.3. Gravimetric Parameters

Bulk density ($p_b$) was calculated as the ratio of the seeds/kernels mass (M/m) and their total occupied volume (V). The method described by Konak et al., 2002 [105], with some modifications, was used.

$$p_b \ (kg/m^3) = M/V \tag{8}$$

True density ($p_t$) was analyzed with toluene displacement method using a pycnometer. The method consists in immersion of a weighted quantity of seeds in toluene ($p$ = 0.867 g/mL) and then recording the volume displaced [40].

$$p_t \left( kg/m^3 \right) = (M \times p_{toluene}) / M_T \tag{9}$$

Porosity was calculated as a function of the previous densities according on the Equation (10) [37,39,40,49].

$$pt \ (\%) = \frac{pt - pb}{pt} \times 100 \tag{10}$$

### 3.4. Comparison of the Seeds, Oil and Oilcakes

#### 3.4.1. Free Amino Acids Determination

Prior to analysis, the sample (3.7 $\pm$ 0.5 g) was mixed with 30 mL of 15% trichloroacetic acid (TCA). The pH of the solution was adjusted to 2.2 with sodium hydroxide solution and further diluted to exactly 50 mL with 15% TCA. After centrifugation for 5 min at 3000 rpm, the supernatant was filtrated through a 0.45 µm nylon filter [106,107]. The solution contains primary and secondary amino acids that were further analyzed with the Ez:faast GC-MS kit (Phenomenex, Torrance, CA, USA). The method was carried out in three steps: a solid phase extraction (performed in sorbent packed tips that bids amino acids and allow the other compounds to flow through), quick derivatization (amino acids migrate to the organic layer) and a liquid/liquid extraction (the organic layer was removed, evaporated and re-dissolved in solvent). Amino acids analysis was made with a gas-chromatograph coupled with a mass spectroscopy instrument (Shimadzu, Kyoto, Japan). The entire time of analysis was 10 min and the injected volume was set at 0.002 mL. The amino acids separation was performed in a ZB-AAA (10 m $\times$ 0.25 mm) column. It was applied the split-less injection mode. The initial temperature of the GC oven was 110 °C, which was increased until 320 °C and held for three minutes. The condition employed for the mass spectrometer were 200 °C for the ion source and 320 °C for the interface. The quadrupole measured the abundance of ions from 35 to 500 m/z. For the calibration, there were used solutions with amino acids mixture included in the kit mentioned above [107].

#### 3.4.2. Fatty Acids Determination

The oils extracted from oilcakes and seeds and the one provided by the local factory were analyzed for fatty acids methyl esters (FAME). Fatty acids derivation was performed according to the following procedure: 0.1 g of each oil sample was mixed with 0.4 mL of n-hexane and 0.4 mL of 15% boron trifluoride in methanol. The solution was heated at 60 °C in a water bath (Memmert Waterbath WNB 22, Schwabach, Germany) for 5 min. After cooling to room temperature, it was mixed with 2 mL of saturated sodium chloride solution and centrifugated at 2000 rpm for three minutes. After that, the supernatant was filtrated through a 0.45 µm nylon filter. FAME's separation was made on a GC-MS instrument (GC MS-QP 2010 Plus, Shimadzu, Kyoto, Japan), using a SUPELCOWAX 10 column (length of 60 m, inner diameter of 0.25 mm ID and 0.25 µm film thickness; Supelco Inc., Bellefonte, PA, USA). The GC oven temperature increased at a rate of 7 °C/min from 140 °C to 220 °C, which was held for 23 min. The flow rate of the carrier gas (He) was kept at 0.8 mLmin$^{-1}$. The injection port temperature was 210 °C. The temperature condition employed for the mass spectrometer were 250 °C for interface and 180 °C for the ion source.

Ion electron impact mass spectra were recorded at a positive ionization energy of 70 eV. The scans range was 22–395 m/z (0.14 scans/s). Under these conditions, a 0.001 mL sample was injected in split mode 1:24. FAME's identification and quantification were performed by comparing their retention times with calibration curves obtained by reference standard FAME's mixed (Restek, Lisses, France) [108].

### 3.4.3. Mineral Estimation

The mineral elements were estimated with a system Agilent Technologies 7500 Series (Agilent Technologies, Santa Clara, CA, USA) coupled-mass spectrometer. The parameters were nebulizer 0.9 mL/min, RF power 1500 W, carrier gas 0.92 L/min, makeup gas 0.17 L/min, mass range 7–205 uma, integration time 0.1 s, acquisition 22.76 s. Detector parameters were: discriminator 8 mV, analogue HV 1770 V and pulse HV 1070 V. Prior to analysis, five grams of sample were calcinated at 550 °C for 6 h. The ash was dissolved with 0.73 mL nitric acid (65% HNO3, Sigma Aldrich, Darmstadt, Germany), placed into a 50 mL flask and completed with deionized water. The elements standard solutions were prepared by diluting stock solution of 1000 mg/L of Li, Be, Mg, Ti, Tl, Co, As, Ca, Cd, Cr, Ce, Cu, Hg, Fe, Mn, Ni, Pb, Se, Sr, Sb, Mo, V and Zn [73].

### 3.5. Statistical Analysis

All results are presented as mean $\pm$ standard deviation. The chemical and functional analyses for sunflower seeds and oilcakes were performed in triplicate. While, amino acids, fatty acids and minerals analyses were performed in duplicate. The physical characteristics of sunflower seeds were performed on a sample of 500 seeds. The values obtained were processed by using SPSS 25.0 (trial version) software (IBM, New York, NY, USA). The difference between samples was established by analysis of variance (ANOVA) using Turkey's test at a 5% significance level. To determine the difference between the two samples of oilcakes a Student's *t*-test was performed.

## 4. Challenges and Future Perspective

Sunflower oilcakes, the by-products obtained after the extraction of oil for sunflower seeds, are rich in dietary fibers and proteins. In the literature, we found numerous studies about the extraction of proteins and their utilization in the food industry, but none about the utilization of dietary fibers. Moreover, there is an increasing demand for an alternative source of plant proteins opening an opportunity for the utilization of sunflower oilcakes as a source of proteins, due to the well-balanced amino acid profile.

Sunflower oilcakes represent also a successful alternative for future food packaging because they are renewable, low-cost resources. Moreover, they permit the development of completely edible and biodegradable materials.

## 5. Conclusions

Nowadays consumers are interested in healthy lifestyles with diets rich in nutrients and for this reason, the consumption of sunflower seeds, with a high nutritive profile, is recommended. Moreover, the current environmental issues caused by the large amount of waste produced by the food industry impose the research of new ways to transform and re-circulate these resources (very rich in bioactive compounds) into the supply chain.

For increasing the interest in food products containing valuable compounds it is necessary to reduce the prices and improve sensorial acceptance. In this sense, they have evaluated the physicochemical and functional properties of sunflower oilcake. The conclusion obtained indicated that sunflower oilcakes represent due to their nutritive properties a good source for good functional products, conferring also good sensorial characteristics.

The high composition in essential amino acids (45.82% for SFOC/PE and 28.4% for SFOC/C), fibers (31.88% and 12.64% for the oilcakes in form of pellets and cake, respectively) and proteins (21.60% for SFOC/C and 20.15 for SFOC/PE) make the oilcakes an interesting option for human consumption. Moreover, they still contain lipids (15.77%

for pellets and 14.16% for cake) rich in unsaturated fatty acids (95.59% and 92.12% for pellets and cake, respectively). The predominant fatty acids are oleic (in pellets 29.32% and in cake 10.34%) and linoleic (32.81 and 65.88% for cake and pellets, respectively) acids which are essential because they cannot be synthesized in the human body. The amino acids profile shows a balanced composition between essential and non-essential (SFOC/PE presents 45.82% and 54.18% essential and non-essential AA respectively). The valorization of these by-products by incorporation in the human diet contributes to the realization of sustainability and circular economy.

Sunflower seeds are very nutritive (33.85% proteins, 65.42% lipids and 18 mineral elements). Due to the rich content in lipids, they are principally used as a source for vegetable oil, which is predominant in kernels (23.73%). To choose high-quality sunflower seeds with the highest fat contents, a classification based on the mass is required. An attentive analysis shows that class III permits obtaining a high oil yield. Further analysis of the extracted oil shows high content in unsaturated fatty acids (93.1%), of which the main are oleic and linoleic acids (respectively, 14.10% and 50.32%). The seeds are a rich source (the value for total AA present in the sample is 5790.26 nmol/g) of essential (6.63%) and non-essential (93.37%) amino acids. On the other hand, the oil extracted from oilcakes and that obtained by cold-pressing present similar chemical properties. Sunflower oil presents essential fatty acids, namely 19.81% oleic acid and 64.35% linoleic acid. Together they help reduce cholesterol levels in the blood and thus prevent heart diseases. A nutritional comparison between oilcakes, seeds and oil shows that they have numerous components suitable for a healthy diet.

**Author Contributions:** A.P., F.U. and S.A. contributed equally to the collection of data and preparation of the paper. All authors have read and agreed to the published version of the manuscript.

**Funding:** This work was supported by a grant from the Ministry of Research, Innovation and Digitization, CNCS/CCCDI–UEFISCDI, project number PN-III-P2-2.1-PED-2019-3863, within PNCDI III.

**Institutional Review Board Statement:** Not applicable.

**Informed Consent Statement:** Not applicable.

**Data Availability Statement:** Not applicable.

**Acknowledgments:** This work was supported by a grant from the Ministry of Research, Innovation and Digitization, CNCS/CCCDI–UEFISCDI, project number PN-III-P2-2.1-PED-2019-3863, within PNCDI III.

**Conflicts of Interest:** The authors declare no conflict of interest.

## References

1. Otles, S.; Despoudi, S.; Bucatariu, C.; Kartal, C. *Valorization, and Sustainability in the Food Industry*; Elsevier Inc.: Amsterdam, The Netherlands, 2015; ISBN 978-0-12-800351-0.
2. Kot, A.M.; Pobiega, K.; Piwowarek, K.; Kieliszek, M.; Błażejak, S.; Gniewosz, M.; Lipińska, E. Biotechnological Methods of Management and Utilization of Potato Industry Waste—A Review. *Potato Res.* **2020**, *63*, 431–447. [CrossRef]
3. Esposito, B.; Sessa, M.R.; Sica, D.; Malandrino, O. Towards circular economy in the agri-food sector. A systematic literature review. *Sustainability* **2020**, *12*, 7401. [CrossRef]
4. Borrello, M.; Caracciolo, F.; Lombardi, A.; Pascucci, S.; Cembalo, L. Consumers' perspective on circular economy strategy for reducing food waste. *Sustainability* **2017**, *9*, 141. [CrossRef]
5. Kowalczewski, P.Ł.; Olejnik, A.; Rybicka, I.; Zielińska-Dawidziak, M.; Białas, W.; Lewandowicz, G. Membrane filtration-assisted enzymatic hydrolysis affects the biological activity of potato juice. *Molecules* **2021**, *26*, 852. [CrossRef] [PubMed]
6. Kieliszek, M.; Piwowarek, K.; Kot, A.M.; Pobiega, K. The aspects of microbial biomass use in the utilization of selected waste from the agro-food industry. *Open Life Sci.* **2020**, *15*, 787–796. [CrossRef] [PubMed]
7. Kowalczewski, P.Ł.; Olejnik, A.; Białas, W.; Kubiak, P.; Siger, A.; Nowicki, M.; Lewandowicz, G. Effect of Thermal Processing on Antioxidant Activity and Cytotoxicity of Waste Potato Juice. *Open Life Sci.* **2019**, *14*, 150–157. [CrossRef] [PubMed]
8. Kumar, S.; Kushwaha, R.; Verma, M.L.; Elsevier, B.V. *Recovery and Utilization of Bioactives from Food Processing Waste*; Elsevier B.V.: Amsterdam, The Netherlands, 2019; ISBN 9780444643230.
9. Ancuţa, P.; Sonia, A. Oil press-cakes and meals valorization through circular economy approaches: A review. *Appl. Sci.* **2020**, *10*, 7432. [CrossRef]

10. Gupta, A.; Sharma, R.; Sharma, S.; Singh, B. Oilseed as Potential Functional Food Ingredient. In *Trends & Prospects in Food Technology, Processing and Preservation*, 1st ed.; Prodyut Kumar, P., Mahawar, M.K., Abobatta, W., Panja, P., Eds.; Today and Tomorrow's Printers and Publishers: New Delhi, India, 2018; pp. 25–58.

11. Adeleke, B.S.; Babalola, O.O. Oilseed crop sunflower (*Helianthus annuus*) as a source of food: Nutritional and health benefits. *Food Sci. Nutr.* **2020**, *8*, 4666–4684. [CrossRef]

12. Chauhan, V. Nutritional quality analysis of sunflower seed cake (SSC). *Pharma Innov. J.* **2021**, *10*, 720–728.

13. Mirpoor, S.F.; Giosafatto, C.V.L.; Porta, R. Biorefining of seed oil cakes as industrial co-streams for production of innovative bioplastics. A review. *Trends Food Sci. Technol.* **2021**, *109*, 259–270. [CrossRef]

14. Gültekin Subaşı, B.; Vahapoğlu, B.; Capanoglu, E.; Mohammadifar, M.A. A review on protein extracts from sunflower cake: Techno-functional properties and promising modification methods. *Crit. Rev. Food Sci. Nutr.* **2021**, 1–16. [CrossRef] [PubMed]

15. Ahmar, S.; Gill, R.A.; Jung, K.H.; Faheem, A.; Qasim, M.U.; Mubeen, M.; Zhou, W. Conventional and molecular techniques from simple breeding to speed breeding in crop plants: Recent advances and future outlook. *Int. J. Mol. Sci.* **2020**, *21*, 2590. [CrossRef] [PubMed]

16. Jocić, S.; Miladinović, D.; Kaya, Y. *Breeding and Genetics of Sunflower*; AOCS Press: Champaign, IL, USA, 2015; ISBN 9781630670627.

17. Pal, D. *Sunflower (Helianthus annuus L.) Seeds in Health and Nutrition*; Elsevier Inc.: Amsterdam, The Netherlands, 2011; ISBN 9780123756886.

18. Romanić, R. Cold Pressed Sunflower (*Helianthus annuus* L.) oil. In *Cold Pressed Oils*; Academic Press: Cambridge, MA, USA, 2020; pp. 197–218. [CrossRef]

19. Islam, R.T.; Hossain, M.M.; Majumder, K.; Tipu, A.H. In vitro Phytochemical Investigation of Helianthus annuus Seeds. *Bangladesh Pharm. J.* **2016**, *19*, 100–105. [CrossRef]

20. Anjum, F.M.; Nadeem, M.; Khan, M.I.; Hussain, S. Nutritional and therapeutic potential of sunflower seeds: A review. *Br. Food J.* **2012**, *114*, 544–552. [CrossRef]

21. Ivanova, P.; Chalova, V.; Koleva, L.; Pishtiyski, I. Amino acid composition and solubility of proteins isolated from sunflower meal produced in Bulgaria. *Int. Food Res. J.* **2013**, *20*, 2995–3000.

22. Sarwar, F. The role of oilseeds nutrition in human health: A critical review. *J. Cereals Oilseeds* **2013**, *4*, 97–100. [CrossRef]

23. Savoire, R.; Lanoisellé, J.L.; Vorobiev, E. Mechanical Continuous Oil Expression from Oilseeds: A Review. *Food Bioprocess Technol.* **2013**, *6*, 1–16. [CrossRef]

24. Ramadan, M.F. *Introduction to Cold Pressed Oils: Green Technology, Bioactive Compounds, Functionality, and Applications*; Elsevier Inc.: Amsterdam, The Netherlands, 2020; ISBN 9780128181881.

25. Poiana, M.; Alexa, E.; Moigradean, D.; Popa, M. The influence of the storage conditions on the oxidative stability and antioxidant properties of sunflower and pumpkin oil. In Proceedings of the 44th Croatian & 4th International Symposium of Agriculture, Opatija, Croatia, 16–20 February 2009; pp. 449–453.

26. Zoumpoulakis, P.; Sinanoglou, V.J.; Siapi, E.; Heropoulos, G.; Proestos, C. Evaluating modern techniques for the extraction and characterisation of sunflower (*Hellianthus annus* L.) seeds phenolics. *Antioxidants* **2017**, *6*, 46. [CrossRef] [PubMed]

27. Avni, T.C.A.; Anupriya, S.; Rai, P.; Maan, K. Effects of Heating and Storage on Nutritional value of Sunflower Oil. *DU J. Undergrad. Res. Innov.* **2016**, *2*, 196–202.

28. Nadeem, M.; Situ, C.; Mahmud, A.; Khalique, A.; Imran, M.; Rahman, F.; Khan, S. Antioxidant activity of sesame (*Sesamum indicum* L.) cake extract for the stabilization of olein based butter. *JAOCS J. Am. Oil Chem. Soc.* **2014**, *91*, 967–977. [CrossRef]

29. Nandha, R.; Singh, H.; Garg, K.; Rani, S. Therapeutic Potential of Sunflower Seeds: An Overview. *Int. J. Res. Dev. Pharm. Life Sci.* **2014**, *3*, 967–972.

30. Bochkarev, M.S.; Egorova, E.Y.; Reznichenko, I.Y.; Poznyakovskiy, V.M. Reasons for the ways of using oilcakes in food industry. *Foods Raw Mater.* **2016**, *4*, 4–12. [CrossRef]

31. Alagawany, M.; Farag, M.R.; El-Hack, M.E.A.; Dhama, K. The Practical Application of Sunflower Meal in Poultry Nutrition. *Adv. Anim. Vet. Sci.* **2015**, *3*, 634–648. [CrossRef]

32. Serrapica, F.; Masucci, F.; Raffrenato, E.; Sannino, M.; Vastolo, A.; Barone, C.M.A.; Di Francia, A. High fiber cakes from mediterranean multipurpose oilseeds as protein sources for ruminants. *Animals* **2019**, *9*, 918. [CrossRef] [PubMed]

33. Nang Thu, T.T.; Bodin, N.; Saeger, S.; Larondelle, Y.; Rollin, X. Substitution of fish meal by sesame oil cake (*Sesamum indicum* L.) in the diet of rainbow trout (*Oncorhynchus mykiss* W.). *Aquac. Nutr.* **2011**, *17*, 80–89. [CrossRef]

34. Wanjari, N.; Waghmare, J. Phenolic and antioxidant potential of sunflower meal. *Pelagia Res. Libr. Adv. Appl. Sci. Res.* **2015**, *6*, 221–229.

35. Lazaro, E.; Benjamin, Y.; Robert, M. The Effects of Dehulling on Physicochemical Properties of Seed Oil and Cake Quality of Sunflower. *Tanzania J. Agric. Sci.* **2014**, *13*, 41–47.

36. Araujo, M.E.V.; Barbosa, E.G.; Gomes, F.A.; Teixeira, I.R.; Lisboa, C.F.; Araújo, R.S.L.; Corrêa, P.C. Physical properties of sesame seeds harvested at different maturation stages and thirds of the plant. *Chil. J. Agric. Res.* **2018**, *78*, 495–502. [CrossRef]

37. Ortiz-Hernandez, A.A.; Araiza-Esquivel, M.; Delgadillo-Ruiz, L.; Ortega-Sigala, J.J.; Durán-Muñoz, H.A.; Mendez-Garcia, V.H.; Yacaman, M.J.; Vega-Carrillo, H.R. Physical characterization of sunflower seeds dehydrated by using electromagnetic induction and low-pressure system. *Innov. Food Sci. Emerg. Technol.* **2020**, *60*, 102285. [CrossRef]

38. Igbozulike, A.O.; Amamgbo, N. Effect of Moisture Content on Physical Properties of Fluted Pumpkin Seeds. *J. Biosyst. Eng.* **2019**, *44*, 69–76. [CrossRef]

39. Malik, M.A.; Saini, C.S. Engineering properties of sunflower seed: Effect of dehulling and moisture content. *Cogent Food Agric.* **2016**, *2*. [CrossRef]

40. Munder, S.; Argyropoulos, D.; Müller, J. Class-based physical properties of air-classified sunflower seeds and kernels. *Biosyst. Eng.* **2017**, *164*, 124–134. [CrossRef]

41. De Figueiredo, A.K.; Baümler, E.; Riccobene, I.C.; Nolasco, S.M. Moisture-dependent engineering properties of sunflower seeds with different structural characteristics. *J. Food Eng.* **2011**, *102*, 58–65. [CrossRef]

42. Costa, C.; Antonucci, F.; Pallottino, F.; Aguzzi, J.; Sun, D.W.; Menesatti, P. Shape Analysis of Agricultural Products: A Review of Recent Research Advances and Potential Application to Computer Vision. *Food Bioprocess Technol.* **2011**, *4*, 673–692. [CrossRef]

43. Mirzabe, A.H.; Khazaei, J.; Chegini, G.R. Physical properties and modeling for sunflower seeds. *Agric. Eng. Int. CIGR J.* **2012**, *14*, 190–202.

44. Seiler, G.J.; Gulya, T.J. *Sunflower: Overview*, 2nd ed.; Elsevier Ltd.: Amsterdam, The Netherlands, 2016; ISBN 9780081005965.

45. Menzel, C. Improvement of starch films for food packaging through a three-principle approach: Antioxidants, cross-linking and reinforcement. *Carbohydr. Polym.* **2020**, *250*, 116828. [CrossRef] [PubMed]

46. Casoni, A.I.; Gutierrez, V.S.; Volpe, M.A. Conversion of sunflower seed hulls, waste from edible oil production, into valuable products. *J. Environ. Chem. Eng.* **2019**, *7*, 102893. [CrossRef]

47. Dobrzaski, B.; Stpniewski, A. Physical Properties of Seeds in Technological Processes. *Adv. Agrophys. Res.* **2013**, *11*, 269–294. [CrossRef]

48. Rodríguez, M.; Nolasco, S.; Izquierdo, N.; Mascheroni, R.; Madrigal, M.S.; Flores, D.C.; Ramos, A.Q. Microwave-assisted extraction of antioxidant compounds from sunflower hulls. *Heat Mass Transf.* **2019**, *55*, 3017–3027. [CrossRef]

49. Çetin, N.; Karaman, K.; Beyzi, E.; Sağlam, C.; Demirel, B. Comparative Evaluation of Some Quality Characteristics of Sunflower Oilseeds (*Helianthus annuus* L.) Through Machine Learning Classifiers. *Food Anal. Methods* **2021**, *14*, 1666–1681. [CrossRef]

50. Muttagi, G.C.; Joshi, N. Physico-chemical composition of selected sunflower seed cultivars. *Int. J. Chem. Stud.* **2020**, *8*, 2095–2100. [CrossRef]

51. Akkaya, M.R. Prediction of fatty acid composition of sunflower seeds by near-infrared reflectance spectroscopy. *J. Food Sci. Technol.* **2018**, *55*, 2318–2325. [CrossRef]

52. Nadeem, M.; Anjum, F.M.; Arshad, M.U.; Hussain, S. Chemical characteristics and antioxidant activity of different sunflower hybrids and their utilization in bread. *Afr. J. Food Sci.* **2010**, *4*, 618–626.

53. Kolláthová, R.; Varga, B.; Ivanišová, E.; Bíro, D.; Rolinec, M.; Juráček, M.; Šimko, M.; Gálik, B. Mineral Profile Analysis of Oilseeds and Their By-Products As Feeding Sources for Animal Nutrition. *Slovak J. Anim. Sci* **2019**, *52*, 9–15.

54. Jafari, S.; Khazaei, J.; Arabhosseini, A.; Massah, J.; Khoshtaghaza, M.H. Study on Mechanical Properties of Sunflower Seeds. *Food Sci. Technol.* **2011**, *14*, 6.

55. Gupta, R.K.; Das, S.K. Physical properties of sunflower seeds. *J. Agric. Eng. Res.* **1997**, *66*, 1–8. [CrossRef]

56. Sumon, M.M.; Tinggi, S.; Ekonomi, I.; Surabaya, P.; Hossain, A. Comparative Study on Physicochemical Composition of Different Genotypes of Sunflower Seed and Mineral Profile of Oil Cake. *Agriculturists* **2021**, *18*, 83–93. [CrossRef]

57. Pawar, V.D.; Patil, J.N.; Sakhale, B.K.; Agarkar, B.S. Studies on selected functional properties of defatted sunflower meal and its high protein products. *J. Food Sci. Technol.* **2001**, *38*, 47–51.

58. Taha, F.S.; Mohamed, G.F.; Mohamed, S.H.; Mohamed, S.S.; Kamil, M.M. Optimization of the Extraction of Total Phenolic Compounds from Sunflower Meal and Evaluation of the Bioactivities of Chosen Extracts. *Am. J. Food Technol.* **2011**, *6*, 1002–1020. [CrossRef]

59. Rosa, P.M.; Antoniassi, R.; Freitas, S.C.; Bizzo, H.R.; Zanotto, D.L.; Oliveira, M.F.; Castiglioni, V.B.R. Chemical composition of brazilian sunflower varieties. *Helia* **2009**, *32*, 145–156. [CrossRef]

60. Žilic, S.; Barac, M.; Pešic, M.; Crevar, M.; Stanojevic, S.; Nišavic, A.; Saratlic, G.; Tolimir, M. Characterization of sunflower seed and kernel proteins. *Helia* **2010**, *33*, 103–114. [CrossRef]

61. Santalla, E.M.; Mascheroni, R.H. Note: Physical Properties of High Oleic Sunflower Seeds. *Food Sci. Technol. Int.* **2003**, *9*, 435–442. [CrossRef]

62. Abdullah, M.H.R.O.; Ch'ng, P.E.; Lim, T.H. Some Physical Properties of Parkia Speciosa Seeds. *Int. Conf. Food Eng. Biotechnol.* **2011**, *9*, 43–47.

63. Krajewska, M.; Ślaska-Grzywna, B.; Andrejko, D. Physical Properties of Seeds of the Selected Oil Plants. *Agric. Eng.* **2016**, *20*, 69–77. [CrossRef]

64. Coşkuner, Y.; Gökbudak, A. Dimensional specific physical properties of fan palm fruits, seeds and seed coats (*Washingtonia robusta*). *Int. Agrophys.* **2016**, *30*, 301–309. [CrossRef]

65. Aviara, N.A.; Gwandzang, M.I.; Haque, M.A. Physical properties of guna seeds. *J. Agric. Eng. Res.* **1999**, *73*, 105–111. [CrossRef]

66. Niveditha, V.R.; Sridhar, K.R.; Balasubramanian, D. Physical and mechanical properties of seeds and kernels of canavalia of coastal sand dunes. *Int. Food Res. J.* **2013**, *20*, 1547–1554.

67. Seifi, M.R.; Alimardani, R. Moisture-Dependent Physical Properties of Sunflower (SHF8190). *Mod. Appl. Sci.* **2010**, *4*, 135–143. [CrossRef]

68. Babić, L.J.; Radojčin, M.; Pavkov, I.; Babić, M. The physical and compressive load properties of sunflower (*Helianthus annuus* L.) fruit. *Helia* **2012**, *35*, 95–112. [CrossRef]

69. Popović, S.; Hromiš, N.; Šuput, D.; Bulut, S.; Romanić, R.; Lazić, V. Valorization of By-Products From the Production of Pressed Edible Oils to Produce Biopolymer Films. In *Cold Pressed Oils*; Academic Press: Cambridge, MA, USA, 2020; pp. 15–30. [CrossRef]
70. Sobczak, P.; Zawislak, K.; Starek, A.; Zukiewicz-Sobczak, W.; Sagan, A.; Zdybel, B.; Andrejko, D. Compaction process as a concept of press-cake production from organic waste. *Sustainability* **2020**, *12*, 1567. [CrossRef]
71. Adesina, S.A. Effect of processing on the proximate composition of sunflower (*Helianthus annuus*) seeds. *Agro-Science* **2019**, *17*, 27. [CrossRef]
72. Rani, R.; Badwaik, L.S. Functional Properties of Oilseed Cakes and Defatted Meals of Mustard, Soybean and Flaxseed. *Waste Biomass Valorization* **2021**, *12*, 5639–5647. [CrossRef]
73. Sinkovič, L.; Kolmanič, A. Elemental composition and nutritional characteristics of cucurbita pepo subsp. Pepo seeds, oil cake and pumpkin oil. *J. Elem.* **2021**, *26*, 97–107. [CrossRef]
74. Sunil, L.; Appaiah, P.; Prasanth Kumar, P.K.; Gopala Krishna, A.G. Preparation of food supplements from oilseed cakes. *J. Food Sci. Technol.* **2015**, *52*, 2998–3005. [CrossRef]
75. Cozea, A.; Ionescu, N.; Popescu, M.; Neagu, M.; Gruia, R. Comparative study concerning the composition of certain oil cakes with phytotherapeutical potential. *Rev. Chim.* **2016**, *67*, 422–425.
76. Hussain, S.; Jõudu, I.; Bhat, R. Dietary fiber from underutilized plant resources-A positive approach for valorization of fruit and vegetable wastes. *Sustainability* **2020**, *12*, 5401. [CrossRef]
77. Dhingra, D.; Michael, M.; Rajput, H.; Patil, R.T. Dietary fibre in foods: A review. *J. Food Sci. Technol.* **2012**, *49*, 255–266. [CrossRef] [PubMed]
78. Maphosa, Y.; Jideani, V.A. Dietary fiber extraction for human nutrition—A review. *Food Rev. Int.* **2016**, *32*, 98–115. [CrossRef]
79. Nevara, G.A.; Kharidah, S.; Muhammad, S.; Zawawi, N.; Mustapha, N.A.; Karim, R. Dietary Fiber: Fractionation, Characterization and Potential Sources from Defatted Oilseeds. *Foods* **2021**, *10*, 754. [CrossRef] [PubMed]
80. Bhise, S.R.; Kaur, A.; Manikantan, M.R.; Singh, B. Development of textured defatted sunflower meal by extrusion using response surface methodology. *Acta Aliment.* **2015**, *44*, 251–258. [CrossRef]
81. Bhise, S.; Kaur, A. The effect of extrusion conditions on the functional properties of defatted cake of sunflower-maize based expanded snacks. *Int. J. Food Ferment. Technol.* **2015**, *5*, 247. [CrossRef]
82. Brachet, M.; Arroyo, J.; Bannelier, C.; Cazals, A.; Fortun-Lamothe, L. Hydration capacity: A new criterion for feed formulation. *Anim. Feed Sci. Technol.* **2015**, *209*, 174–185. [CrossRef]
83. Lin, M.J.; Humbert, E.S.; Sosulski, F.W. Functional Properties Sunflower Meal Products of. *J. Food Sci.* **1974**, *39*, 368–370. [CrossRef]
84. Sosulski, F.; Fleming, S.E. Chemical, functional, and nutritional properties of sunflower protein products. *J. Am. Oil Chem. Soc.* **1977**, *54*, A100. [CrossRef]
85. Grasso, S.; Omoarukhe, E.; Wen, X.; Papoutsis, K.; Methven, L. The use of upcycled defatted sunflower seed flour as a functional ingredient in biscuits. *Foods* **2019**, *8*, 305. [CrossRef]
86. Amza, T.; Amadou, I.; Zhu, K.X.; Zhou, H.M. Effect of extraction and isolation on physicochemical and functional properties of an underutilized seed protein: Gingerbread plum (Neocarya macrophylla). *Food Res. Int.* **2011**, *44*, 2843–2850. [CrossRef]
87. White, N.D.G.; Jayas, D.S. Physical properties of canola and sunflower meal pellets. *Can. Biosyst. Eng.* **2001**, *43*, 349–352.
88. Dabbour, M.; He, R.; Ma, H.; Musa, A. Optimization of ultrasound assisted extraction of protein from sunflower meal and its physicochemical and functional properties. *J. Food Process Eng.* **2018**, *41*, e12799. [CrossRef]
89. Cai, T.; Chang, K.C.; Lunde, H. Physicochemical Properties and Yields of Sunflower Protein Enzymatic Hydrolysates As Affected by Enzyme and Defatted Sunflower Meal. *J. Agric. Food Chem.* **1996**, *44*, 3500–3506. [CrossRef]
90. Melo, D.; Álvarez-Ortí, M.; Nunes, M.A.; Costa, A.S.G.; Machado, S.; Alves, R.C.; Pardo, J.E.; Oliveira, M.B.P.P. Whole or defatted sesame seeds (*Sesamum indicum* L.)? The effect of cold pressing on oil and cake quality. *Foods* **2021**, *10*, 2108. [CrossRef] [PubMed]
91. Arrutia, F.; Binner, E.; Williams, P.; Waldron, K.W. Oilseeds beyond oil: Press cakes and meals supplying global protein requirements. *Trends Food Sci. Technol.* **2020**, *100*, 88–102. [CrossRef]
92. Jannathulla, R.; Dayal, J.S.; Ambasankar, K.; Muralidhar, M. Effect of Aspergillus niger fermented soybean meal and sunflower oil cake on growth, carcass composition and haemolymph indices in Penaeus vannamei Boone, 1931. *Aquaculture* **2018**, *486*, 1–8. [CrossRef]
93. Zentek, J.; Knorr, F.; Mader, A. *Reducing Waste in Fresh Produce Processing and Households through Use of Waste as Animal Feed*; Woodhead Publishing Limited: Cambridge, UK, 2013; ISBN 9781782420187.
94. Soetan, K.O.; Olaiya, C.O.; Oyewole, O.E. The importance of mineral elements for humans, domestic animals and plants: A review. *Afr. J. Food Sci.* **2010**, *4*, 200–222.
95. Njuguna, D.G.; Wanyoko, J.K.; Kinyanjui, T.; Wachira, F.N. Mineral Elements in the Kenyan Tea Seed Oil Cake. *Int. J. Res. Chem. Environ.* **2013**, *3*, 253–261.
96. Chaves, E.S.; dos Santos, E.J.; Araujo, R.G.O.; Oliveira, J.V.; Frescura, V.L.A.; Curtius, A.J. Metals and phosphorus determination in vegetable seeds used in the production of biodiesel by ICP OES and ICP-MS. *Microchem. J.* **2010**, *96*, 71–76. [CrossRef]
97. Goiri, I.; Zubiria, I.; Benhissi, H.; Atxaerandio, R.; Ruiz, R.; Mandaluniz, N.; Garcia-Rodriguez, A. Use of cold-pressed sunflower cake in the concentrate as a low-input local strategy to modify the milk fatty acid profile of dairy cows. *Animals* **2019**, *9*, 803. [CrossRef]

98. Zubiria, I.; Garcia-Rodriguez, A.; Atxaerandio, R.; Ruiz, R.; Benhissi, H.; Mandaluniz, N.; Lavín, J.L.; Abecia, L.; Goiri, I. Effect of feeding cold-pressed sunflower cake on ruminal fermentation, lipid metabolism and bacterial community in dairy cows. *Animals* **2019**, *9*, 755. [CrossRef] [PubMed]

99. Hansen, J.Ø.; Skrede, A.; Mydland, L.T.; Øverland, M. Fractionation of rapeseed meal by milling, sieving and air classification—Effect on crude protein, amino acids and fiber content and digestibility. *Anim. Feed Sci. Technol.* **2017**, *230*, 143–153. [CrossRef]

100. Chetrariu, A.; Dabija, A. Quality Characteristics of Spelt Pasta Enriched with Spent Grain. *Agronomy* **2021**, *11*, 1824. [CrossRef]

101. Omowaye-Taiwo, O.A.; Fagbemi, T.N.; Ogunbusola, E.M.; Badejo, A.A. Effect of germination and fermentation on the proximate composition and functional properties of full-fat and defatted cucumeropsis mannii seed flours. *J. Food Sci. Technol.* **2015**, *52*, 5257–5263. [CrossRef] [PubMed]

102. Onipe, O.O.; Beswa, D.; Jideani, A.I.O. Effect of size reduction on colour, hydration and rheological properties of wheat bran. *Food Sci. Technol.* **2017**, *37*, 389–396. [CrossRef]

103. Coțovanu, I.; Batariuc, A.; Mironeasa, S. Characterization of quinoa seeds milling fractions and their effect on the rheological properties of wheat flour dough. *Appl. Sci.* **2020**, *10*, 7225. [CrossRef]

104. Iyenagbe, D.O.; Malomo, S.A.; Idowu, A.O.; Badejo, A.A.; Fagbemi, T.N. Effects of thermal processing on the nutritional and functional properties of defatted conophor nut (Tetracarpidium conophorum) flour and protein isolates. *Food Sci. Nutr.* **2017**, *5*, 1170–1178. [CrossRef] [PubMed]

105. Konak, M.; Çarman, K.; Aydin, C. Physical properties of chick pea seeds. *Biosyst. Eng.* **2002**, *82*, 73–78. [CrossRef]

106. Dabadé, D.S.; Jacxsens, L.; Miclotte, L.; Abatih, E.; Devlieghere, F.; De Meulenaer, B. Survey of multiple biogenic amines and correlation to microbiological quality and free amino acids in foods. *Food Control* **2021**, *120*, 107497. [CrossRef]

107. Gifty, A.G.; De Meulenaer, B.; Olango, T.M. Variation in tuber proximate composition, sugars, fatty acids and amino acids of eight Oromo dinich (Plectranthus edulis) landraces experimentally grown in Ethiopia. *J. Food Compos. Anal.* **2018**, *67*, 191–200. [CrossRef]

108. Dulf, F.V. Fatty acids in berry lipids of six sea buckthorn (*Hippophae rhamnoides* L., subspecies carpatica) cultivars grown in Romania. *Chem. Cent. J.* **2012**, *6*, 106. [CrossRef] [PubMed]

*Review*

# Breeding Canola (*Brassica napus* L.) for Protein in Feed and Food

Kenny K. Y. So and Robert W. Duncan *

Department of Plant Science, University of Manitoba, Winnipeg, MB R3T 2N2, Canada; sok@myumanitoba.ca
* Correspondence: rob.duncan@umanitoba.ca

**Abstract:** Interest in canola (*Brassica napus* L.). In response to this interest, scientists have been tasked with altering and optimizing the protein production chain to ensure canola proteins are safe for consumption and economical to produce. Specifically, the role of plant breeders in developing suitable varieties with the necessary protein profiles is crucial to this interdisciplinary endeavour. In this article, we aim to provide an overarching review of the canola protein chain from the perspective of a plant breeder, spanning from the genetic regulation of seed storage proteins in the crop to advancements of novel breeding technologies and their application in improving protein quality in canola. A review on the current uses of canola meal in animal husbandry is presented to underscore potential limitations for the consumption of canola meal in mammals. General discussions on the allergenic potential of canola proteins and the regulation of novel food products are provided to highlight some of the challenges that will be encountered on the road to commercialization and general acceptance of canola protein as a dietary protein source.

**Keywords:** canola; rapeseed; *Brassica napus*; canola protein; plant proteins; breeding; food safety

## 1. Introduction

The global population is projected by the United Nations to increase beyond nine billion by the mid-21st century (Figure 1) [1] and scientists have been tasked with ensuring food security to sustain this growth [2–4]. Agriculture will be able to feed the global population provided improvements are made to the sustainability of current agricultural practices along with a concomitant shift in dietary preferences [4–7]. Complicating the task of food production are increasingly severe and unpredictable climatic patterns [6,8,9] as well as the continued reduction in arable land [6,8–10]. The concept of food security was initially defined by the United Nations at the World Food Conference of 1974 as the availability of food at reasonable prices at all times [11]. The definition of food security has since been expanded to encompass the nutritional and social aspects of food [12,13]. Dietary protein has been a focus of nutrition programs as it is often a limiting macronutrient in malnourished and food-insecure populations [14–16].

Dietary protein is primarily acquired through either the consumption of meat (animal protein) or legumes. Historically, meat consumption was associated with economic wealth and recent shifts in developing economies have dramatically increased meat consumption [17–19]. Intensive animal husbandry for meat production has been repeatedly cited as being environmentally destructive, unsustainable, and an inefficient method of converting protein feedstock into dietary protein [20–23]. The production of plant-based protein sources, such as kidney bean (*Phaseolus vulgaris* L.), requires substantially less input and generates less waste than an equivalent unit of red meat such as beef [20]. Given the large amount of feed required to produce animal protein [6], the redistribution of land from feed grain to crop production enables larger quantities of dietary protein to be produced per unit area [9,24]. Evidently, even partial replacement of animal proteins with plant-based alternatives will have considerable impact on long-term sustainability [25].

Although plant-based proteins are prominent in many cultures, European and North American diets show a proclivity for animal proteins in their diet [16,17,26]. However, a

**Citation:** So, K.K.Y.; Duncan, R.W. Breeding Canola (*Brassica napus* L.) for Protein in Feed and Food. *Plants* **2021**, *10*, 2220. https://doi.org/10.3390/plants10102220

Academic Editor: Ivo Vaz de Oliveira

Received: 30 August 2021
Accepted: 11 October 2021
Published: 19 October 2021

**Publisher's Note:** MDPI stays neutral with regard to jurisdictional claims in published maps and institutional affiliations.

recent assessment of food supplies across 171 countries found a decrease in animal protein supply in six countries across North America, Europe, and Australia [27], suggesting a gradual replacement of animal protein with alternative proteins such as those from plants. Recent work has demonstrated that the composition of the human gut microbiome is in rapid flux with changes in diet [28], suggesting that humans are amenable to adapting to new diets with relative ease. The challenge of adopting plant-based proteins, therefore, may largely be cultural. In an effort to promote vegetable proteins as a nutritious, sustainable, and a secure alternative to animal protein, 2016 was designated as the International Year of Pulses [29]. Consumer food choices are influenced by dietary guidelines published by national food authorities, and different recommendations show varying degrees of sustainability based on the resources required for production [30]. Given their far-reaching influence, incorporating considerations for sustainability in government food recommendations may substantially affect food security for many people [31]. In Canada, a recent revision to the food guide changing a "meat and alternatives" food group to "protein foods" with an emphasis on plant-based proteins reinforced the suitability of plant proteins as a replacement for animal proteins [32]. In expanding the available sources of plant-based proteins, rapeseed/canola (*Brassica napus* L.) protein has been suggested as a possible plant-based protein source. With increasing global acreages annually (Figure 2), canola that is currently grown and processed for edible oil production generates a large quantity of protein-rich seed meal as a byproduct of the oil extraction process [33]. The ease of access to and availability of canola meal makes it a logically suitable and sustainable candidate for development into a plant-based protein for human consumption [34]. Such a proposal has been supported by both industry and public institutions worldwide, as reflected by numerous abstracts at the latest International Rapeseed Congress [35].

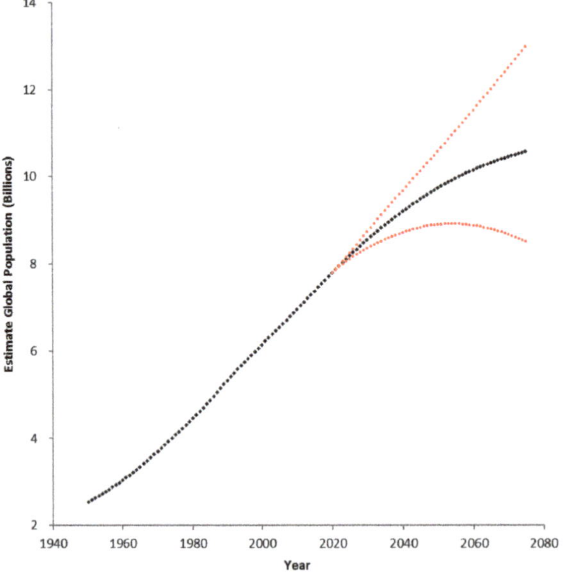

**Figure 1.** The world population is projected to reach nine billion by the mid-21st century. Data points up to 2020 are annual estimates. Black data points from 2021 to 2075 are estimates based on median fertility. Red data trends indicate high- and low-fertility variant estimates (raw data source: United Nations, Department of Economic and Social Affairs, Population Division (2019). World Population Prospects 2019, Online Edition. Rev. 1, accessed on 1 September 2021.

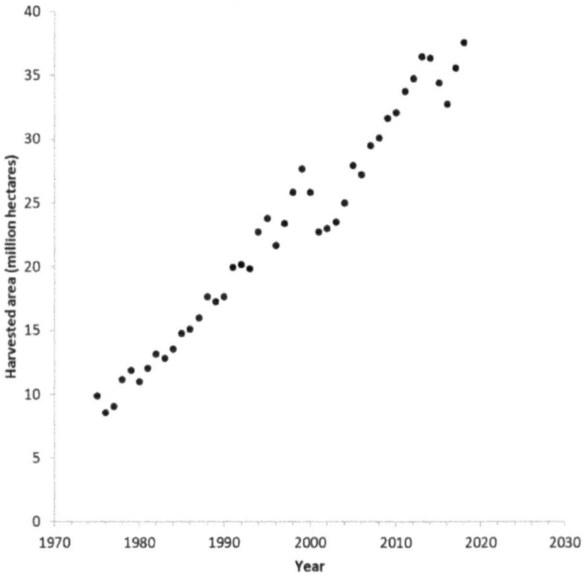

**Figure 2.** Global rapeseed (*Brassica napus* L.) harvested hectares have increased annually since the introduction of the first low-erucic acid, low-glucosinolate canola cultivar 'Tower' in 1974 (data source: FAOSTAT, Food and Agriculture Organization of the United Nations, accessed on 1 September 2021).

In addition to serving as a dietary protein supplement, the functional properties of canola protein render them useful in a diverse range of food processing applications [33,34,36,37]. In such a case, individual canola proteins with the desired functionality would need to be isolated with additional processing steps from the crude seed protein given the differences in functional properties of each protein constituent [34,38–40].

## 2. Transition from Rapeseed to Canola

Rapeseed is an economically important allotetraploid oilseed crop derived from *B. oleracea* L. and *B. rapa* L. [41]. Rapeseed was introduced into Canada in 1943 initially as a source of vegetable oil for industrial applications [42,43]. Despite the high oil content of rapeseed and its adaptability to be grown in Canada, the high percentage of erucic acid in rapeseed oil (approximately 40%) and high levels of glucosinolates (80 µmol/g seed) rendered the oil unfit for the human diet [44,45]. Erucic acid consumption was implicated in the cause of multiple pathologies in various animal models and thus its elimination was a prerequisite for the adoption of rapeseed oil for human consumption [46,47]. Glucosinolates are generally deemed innocuous when intact; however, their degradation products have been implicated as the causal agent of various health problems and are pungent, rendering both the oil and meal unpalatable [48]. Specifically, studies in rats repeatedly showed glucosinolates to adversely affect thyroid function through antagonism against iodine [49,50]. Thus, the elimination of glucosinolates was necessary before rapeseed oil could be incorporated into the human diet [51]. Canadian breeding efforts led to the development of the first low-erucic-acid and low-glucosinolate rapeseed cultivar, named Tower [52]. This and subsequent low-erucic acid, low-glucosinolate cultivars were termed "canola" in North America, and double-low or 00 rapeseed in Europe, to distinguish them from the industrial rapeseed [53]. Current commercial canola cultivars have near undetectable levels of glucosinolates (less than 20 µmol/g seed) and typically have less than 1% erucic acid, well below the maximum allowable levels of 2% erucic acid and 30 µmol glucosinolate/g seed as established by international law [53,54]. The near-elimination of erucic acid and glucosinolates deemed canola to be generally recognized as safe by the United States Food and

Drug Administration (Direct Food Substances Affirmed as Generally Recognized as Safe, 21 C.F.R. §184.1555, [50 FR 3755, Jan. 28, 1985]) and is the fourth largest oilseed produced globally behind oil palm, soybean, and seed cotton (Figure 3) [55]. Domestically, canola production in Canada has grown steadily since its introduction in 1974 to approximately 20 million metric tonnes (Figure 4) and was estimated to contribute just under $30 billion to the Canadian economy [56].

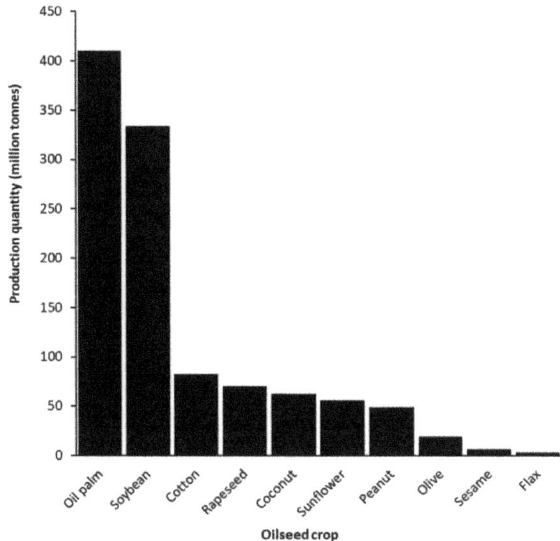


**Figure 3.** Rapeseed (including canola) was the fourth largest oil crop in 2019 based on global production quantity (data source: FAOSTAT, Food and Agriculture Organization of the United Nations, accessed January 2021).

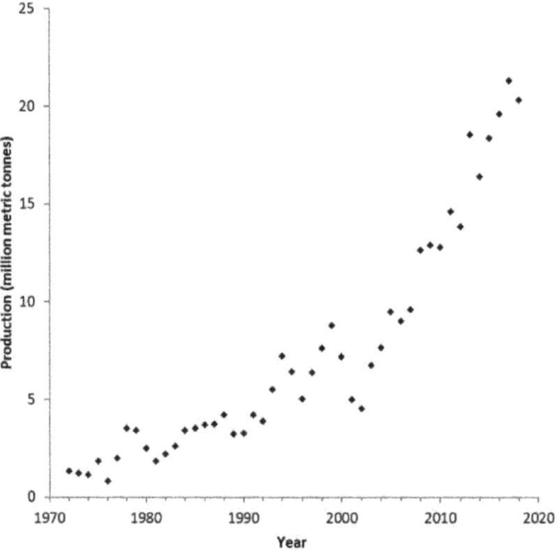


**Figure 4.** Canola production in Canada has increased steadily since the introduction of the first low-erucic acid, low-glucosinolate canola cultivar 'Tower' in 1974 (Source: Statistics Canada. Table 32-10-0359-01, accessed on 1 September 2021).

Though unfit for human consumption, the potential value of erucic acid as an industrial lubricant and raw material for chemical manufacturing was realized by the USDA in the late 1960s [43,57]. At the same time, high-erucic-acid rapeseed (HEAR) cultivars with low glucosinolate levels were developed in Sweden and HEAR breeding genotypes were developed in Canada [58]. The first Canadian HEAR cultivar 'Reston' was registered by the University of Manitoba in 1982 (Registration number 2190).

Although canola remains primarily a crop grown for oil, its potential value as a renewable and sustainable protein source has been gaining attention. Whereas extensive breeding efforts continue to focus on improving the quality and quantity of oil in canola [59], minimal attention has been directed to improving protein quality and content due to their inverse relationship with oil content [60,61]. A canola cultivar with improved meal quality was assigned a patent to Agrigenetics Inc. and Dow AgroSciences LLC in 2016 claiming 45% crude seed protein and less than 18% fibre (US20120213909A1). To better understand the challenges of developing canola protein as a dietary protein source, we need to first consider the technical, biochemical, and genetic factors that affect proteins in canola seed.

## 3. Extracting Oil from Raw Seed

Canola/rapeseed meal is the residual portion of the seed that remains after the oil has been extracted [34]. The oil extraction process is largely divided into a pre-processing step and an extraction step [62]. Pre-processing ensures maximum oil recovery from the seed: the canola seeds are pre-heated, after which they are flattened by rollers into flakes, and subsequently cooked [63–65]. Heating the seeds increases the pliability of the seed to ensure they can be thoroughly flaked without shattering while the flaking and cooking steps function primarily to rupture the seed to allow oil to be released and to increase the surface area of the seed [65].

The extraction step begins with the physical removal of oil from the cooked flakes by screw press [62,65]. Remaining oil in the seed cake is then repeatedly extracted using a mixture of solvents collectively known as isohexane [65]. The meal and oil diverge in their processing after this solvent extraction; the oil is diverted for refining while the seed cake is sent to a desolventizer toaster, where a combination of steam and heat remove residual solvent from the cake [62,65,66]. The resulting oil-free meal is then typically pressed into pellets for animal feed. Alternatively, protein can be isolated from the meal. The key determinate used to differentiate canola from rapeseed is its oil quality, and their oil-free meals differ in glucosinolate content; however, such differentiation cannot be made for purified protein products from either crop. Henceforth, all protein products that are purified from meal originating from canola and rapeseed will be referred to as canola protein. The potential uses of canola protein as a food processing additive as well as non-food, non-feed applications of canola proteins have been recently reviewed [33].

## 4. Effects of Processing on Meal Quality

The quality of canola meal destined for animal feed is judged largely on its amino acid profile and digestibility [67,68]. Conversely, the quality of canola protein destined for food processing is judged on its technical functionality [37,69,70], which is conferred by the molecular structure and content of the individual proteins contained within; thus, the quality of canola protein relates to the integrity of its components.

Extraction conditions can be altered to selectively extract different protein products with different purities and technical functionalities from canola meal [40,71–73] and multiple patents have been assigned to different institutes to protect the intellectual property rights associated with these optimized conditions [64]. Harsh extraction processes can compromise protein quality: while seed proteins can be denatured by exposure to solvents [74], high temperatures alters the digestibility of canola meal and affects the structure of individual canola proteins [48,72,75–77]. Structural changes in canola proteins can abate their functional properties [78,79]. As a feed supplement, protein content is routinely determined by its nitrogen content through combustion analysis (ISO 16634-1:2008; AOAC

Method 990.03) thus, denatured proteins should not affect the protein determination of the meal. However, prolonged desolventizing can negatively impact the bioavailability of essential amino acids such as lysine, due to their modification from Maillard reactions [72,80]. In addition, the digestibility of canola meal decreases with prolonged toasting, rendering it less nutritious as a feed supplement [48,75–77]. If the meal protein is to be used as a functional ingredient in food processing, it is crucial that the structures of the proteins are not compromised.

## 5. Nutritional Value of Canola Meal Protein

Despite the potential reduction in quality during processing, canola meal and protein both have a nutritional profile that is comparable with soy, a common plant protein source. When considering the meal as a whole, the protein content of oil-free canola meal is approximately 38% [81], compared to 44% for its soy counterpart [82] (Table 1). Canola meal tends to have high fibre, resulting in lower digestibility in animals which makes the meal less competitive compared to soy meal models [83,84]. Currently, the majority of canola meal is used for animal feed; however, isolated protein products have the potential to be used as a dietary protein source for humans [33].

**Table 1.** Comparison of seed quality between Canadian soybeans and western Canadian canola presented as the average value of 2013–2017 crops, inclusive. Table compiled with soybean data [82] and canola data [81] published by the Canadian Grain Commission. Non-applicable measurements indicated with n.a.

|  | Soybean | Canola |
|---|---|---|
| Protein content (%) [1] | 34.50 | 20.20 |
| Oil content (%) [1] | 18.40 | 44.50 |
| Oil-free protein of the meal (%) [2] | 43.80 | 37.80 |
| Oleic acid (% in oil) | 21.90 | 62.90 |
| Linoleic acid (% in oil) | 53.60 | 18.70 |
| α-Linolenic acid (% in oil) | 9.00 | 9.40 |
| Total saturated fatty acids (% in oil) | 15.20 | 6.70 |
| Erucic acid (% in oil) | n.a. | 0.01 |
| Total seed glucosinolates (μmol/g, 8.5% moisture) | n.a. | 10.00 |
| Total glucosinolates of the meal (μmol/g, oil-free, dry basis) | n.a. | 21.00 |

[1] Based on 13% moisture in soybean and 8.5% moisture in canola. [2] Based on 13% moisture in soybean and 12% moisture in canola.

Protein quality for human nutrition is most commonly measured with a Protein Digestibility Corrected Amino Acid Score (PDCAAS) [85–87] or, more recently, with a Digestible Indispensable Amino Acid Score [67]. The PDCAAS for a given protein is generally a ratio of its amino acid composition relative to that of a reference protein, then normalized to its digestibility; effectively, a maximum score of 1.0, indicating that one unit of the protein in question is able to supply all the essential amino acids after digestion [85,87]. When considering only the protein fraction, canola protein is comparable with its soy counterpart in its amino acid profile [68] (Table 2) and can generally satisfy human dietary requirements for essential amino acids [33,34]. The postprandial response to canola protein was empirically demonstrated to be equivalent to that of milk protein [88] and soy protein [89].

**Table 2.** Comparison of moisture content, crude protein, and amino acid profiles between soybean meal and western Canadian canola meal. Values presented for canola meal are the mean and standard deviations of samples collected from five different crushing plants across western Canada. The origin and variability of the soybean meal were not disclosed [68].

|  | **Soybean** | **Canola** |
|---|---|---|
| Quality (%) |  |  |
| Moisture | 9.65 | 8.89 ± 0.43 |
| Crude protein | 46.10 | 37.82 ± 2.09 |
| Amino acids (%) |  |  |
| Alanine | 1.97 | 1.64 ± 0.08 |
| Arginine | 3.27 | 2.15 ± 0.11 |
| Aspartic acid | 5.04 | 2.51 ± 0.12 |
| Cysteine | 0.61 | 0.81 ± 0.04 |
| Glutamic acid | 8.34 | 6.2 ± 0.41 |
| Glycine | 1.98 | 1.87 ± 0.10 |
| Histidine | 1.11 | 0.91 ± 0.05 |
| Isoleucine | 1.81 | 1.26 ± 0.09 |
| Leucine | 3.45 | 2.54 ± 0.14 |
| Lysine | 2.90 | 2.09 ± 0.08 |
| Methionine | 0.58 | 0.69 ± 0.03 |
| Phenylalanine | 2.25 | 1.44 ± 0.07 |
| Proline | 2.39 | 2.21 ± 0.11 |
| Serine | 2.31 | 1.47 ± 0.09 |
| Threonine | 1.73 | 1.52 ± 0.06 |
| Tryptophan | 0.69 | 0.52 ± 0.03 |
| Tyrosine | 1.74 | 1.06 ± 0.04 |
| Valine | 1.88 | 1.63 ± 0.09 |

## 6. Anti-Nutrients in Canola Meal

The major drawback of canola protein nutrition is the presence of anti-nutritive compounds that negatively impact health, protein digestion, and amino acid availability [90]. The presence of glucosinolates, phytic acid, sinapine, and tannins reduce the nutritional value of canola meal [90,91].

Glucosinolates are sugar–amino acid conjugates and the content of the aliphatic class of these molecules was successfully reduced during the development of canola; however, residual quantities of phenolic glucosinolates in the meal can be broken down into off-tasting compounds that affect thyroid function [51] and show similar bioactivity to dioxins [92]. Phytic acid (phytates) is the main phosphorus-storage compound in seeds and functions as an antinutrient by binding mineral nutrients and inhibiting digestive enzymes [92]. Sinapine is an unpalatable phenolic compound that functions as an antinutrient by promoting feed-avoidance in sensitive animals [92,93]. Tannins are polyphenolic compounds which, by interacting with proteins and the gastrointestinal tract, reduce the overall digestibility of proteins and bioavailability of amino acids [90]. It should be noted that the degree of sensitivity to these compounds varies between monogastric and ruminant animals.

Progress in traditional breeding and biotechnology have led to the reduction [92–94] and sometimes near elimination of antinutritive compounds such as in the case of phytic acid [95]. For isolated protein products, anti-nutritive compounds are not typically considered a limitation of use as during the process of protein isolation anti-nutritive factors are excluded [96].

## 7. Seed Development in Canola

Prior to examining seed storage proteins, a general understanding of seed development is necessary. A canola seed consists of an embryo and endosperm encased in a seed coat [97]. Seed development in angiosperms is well characterized primarily based on empirical work using the model plant *Arabidopsis thaliana* (L.) Heynh. Seed development has been extensively reviewed in detail by numerous authors and will only be succinctly summarized below. Seed development in canola is broadly divided into three phases: the initial phase where rapid cell division occurs but seed growth is slow; the second phase where cells expand quickly and accumulate both storage protein and lipids; and the third phase where the seed matures and desiccates [98,99].

Embryo development begins with successful double fertilization: the fusion of sperm from the pollen with the egg forms the zygote while the endosperm results from the fusion of a second sperm with the polar nuclei [100]. The zygote then undergoes a series of coordinated divisions to establish polarity and generate tissue layers [101,102]. Embryo developmental stages are described based on their visual morphology; the stages of embryogenesis in canola have been recorded in detailed drawings from light microscopy [103,104] and scanning electron microscopy [105]. The initial globular embryo becomes heart shaped upon initiation of the cotyledon primordia; subsequently the embryo elongates into a torpedo shape; and as the cotyledons develop, they eventually bend over the embryo forming the bent cotyledon stage [103,104,106]. Although the progression from the globular stage to the bent cotyledon stage is ubiquitous across all studied canola genotypes, it is worth noting that the time of occurrence and duration of each stage shows genotypic variation [98,105–108] (Table 3).

Based on Norton and Harris [99], the first phase of seed development is characterized by expeditious growth of the hull (silique) and slow seed growth; this phase of seed development persisted for up to four weeks (28 days) after pollination when examining field-grown *B. napus* plants. When grown under controlled environments, developing seeds of the spring canola cultivar 'Tower' complete the first phase of seed development at approximately 23 days after pollination [98]. The second phase of seed development as described by Norton and Harris [99] occurs five to six weeks post-pollination. This phase is characterized by the initiation of embryo development and the aggregation of storage compounds within the seed. However, light microscopy [107] and scanning microscopy [105] experiments have shown embryo formation to initiate two to three days post-pollination, with globular embryos visible one week after pollination, and torpedo-shaped embryos visible 11–12 days after pollination; Norton and Harris [99] were likely referring to the expansion of the whole seed rather than the embryo. It is during this second phase of seed development that storage proteins begin to accumulate and continue to do so into the third phase of seed development at approximately 6–12 weeks after pollination [98,99].

**Table 3.** Coincidence of zygotic embryogenesis stages in spring *Brassica napus* L. with spatial-temporal patterns in the accumulation of seed storage protein transcripts and gene products during the first six weeks seed of development under controlled growth environment conditions. Abbreviations are as follows: G, globular stage; H, heart stage; T, torpedo stage; C, cotyledon stage; CRU, cruciferin. A single asterisk (*) indicates initial detection and three asterisks (***) indicate peak.

| Weeks after Pollination | 1 | 2 | 3 | 4 | 5 | 6 |
|---|---|---|---|---|---|---|
| **Development** | | | | | | |
| Crouch and Sussex (1981) Planta 153: 64–74 | | | H  T | C | | |
| Custers et al. (1999) Protoplasma 208: 257–264 | G | H  T | C | | | |
| Fernandez et al. (1991) Development 111: 299–313 | | | H | T | C | |
| Ilic-Grubor et al. (1998) Ann. Bot. 82: 157–165 | | H  T  C | | | | |
| Yeung et al. (1996) Int. J. Plant Sci. 157(1): 27–39 | | H  T | C | | | |
| **CRU mRNA** | | | | | | |
| DeLisle and Crouch (1989) Plant Physiol. 91: 617–623 | | | | * | | *** |
| Finkelstein et al. (1985) Plant Physiol. 78: 630–636 | | | * | | | *** |
| Sjodahl et al. (1993) Plant Mol. Biol. 23: 1165–1176 | | | | * | *** | |
| **CRU protein** | | | | | | |
| Crouch and Sussex (1981) Planta 153: 64–74 | | | | * | *** | |
| **NAPIN mRNA** | | | | | | |
| DeLisle and Crouch (1989) Plant Physiol. 91: 617–623 | | | | * | *** | |
| Finkelstein et al. (1985) Plant Physiol. 78: 630–636 | | | * | *** | | |
| **NAPIN protein** | | | | | | |
| Crouch and Sussex (1981) Planta 153: 64–74 | | | | * | *** | |

## 8. Seed Storage Proteins of the Brassicaceae Family

Seed storage proteins (SSP) are a unique class of proteins that are specifically expressed in the developing seed and accumulated within protein storage vacuoles in the mature seed [109,110], though contemporary evidence of synthesis during germination has been reported [111]. These proteins function as a nutrient reservoir by converting nitrogen into storage-stable proteins, which are mobilized during seed germination to support early seedling growth [112,113]. Recently, SSP transcripts have been transiently detected in vegetative tissues in canola when plants were grown under different nitrogen fertility conditions [114] which corroborates the role of these proteins in nitrogen storage.

In the Brassicaceae, the seed protein pool is dominated by the globulin-type SSP cruciferin and the albumin-type SSP napin, along with minor quantities of proteins that play a diverse array of metabolic and physiological roles, respectively [110]. The globulin and albumin designation refers to the classical work of Osborne [115], who differentiated plant proteins based on their differential solubility in various solvents while the numeric values denote the sedimentation coefficient value of the protein. Late embryogenesis abundant (LEA) proteins, for example, are present in the seeds of most plants and function to mediate dehydration tolerance during seed maturation [116]. Specific to the seed protein pool of oilseed species may be substantial quantities of oleosins, proteins that enhance the stability of oil bodies in which these plants accumulate oil [117]. Extensive proteomic studies on the seeds of *B. napus* at various developmental stages have systematized the profusion of proteins present in the seed protein pool [117–120]. Furthermore, these works report remarkable plasticity in the species composition of the seed protein pool through various environmental stimuli [119,121] and physiological processes [118,120]. The spatial-temporal distribution patterns of SSP synthesis and accumulation in *B. napus* during seed development have been described in detail and exhibit genotypic variation [98,121–123].

### 8.1. Cruciferin of B. napus

Cruciferin is a salt-soluble globulin-type protein that accounts for approximately 60% of the total protein pool in canola seed [98]. Related 11S/12S globulin-type SSP from other crop species include cucurbitin from pumpkin (*Cucurbita maxima* Duchesne) and glycinin from soybean (*Glycine max* (L.) Merr.), with the latter being the most similar to cruciferin [110,124]. Mature cruciferin is a large hexameric protein whose subunits are each comprised of a disulfide-bridged $\alpha$ and $\beta$ subunit. Recent work has suggested that cruciferin subunits may be reassembled into a octameric structure for storage inside protein storage vacuoles [125]. The general biosynthesis and deposition of globulin-type SSP have been extensively reviewed [126] and is summarized below.

Globulins are encoded by multigene families. Specifically, in canola, cruciferin is encoded by 9–12 genes [110]. A search on the Uniprot knowledgebase reveals five curated accessions of *B. napus* cruciferin (Table 4) (https://www.uniprot.org, accessed November 2019). Globulin-encoding genes are translated into pre-propolypeptides on the rough endoplasmic reticulum (rER) and at minimum consist of three elements (listed from N to C terminus): a signal peptide, the $\alpha$ subunit, and the $\beta$ subunit [125,126]. The pre-propolypeptide is simultaneously translated and transported into the lumen of the rER and the signal peptide is detached, resulting in a propolypeptide [126]. As the propolypeptide is shuttled through the lumen of the rER, post-translational modifications such as glycosylation occurs followed by the formation of disulfide bridges [126]. The polypeptides are then oligomerized into entropically-favourable trimeric structures that transit to the Golgi apparatus [124,126]. After sorting into vesicles at the trans-Golgi, the trimeric structures are transported to storage protein vacuoles where the prepolypeptides are enzymatically processed by endopeptidases to yield trimers of mature globulin subunits, each consisting of an $\alpha$ and $\beta$ chain, respectively [126,127]. Only upon production of the mature globulin subunits can mature hexameric globulins form [127]. Across different canola genotypes, variation in the temporal accumulation of both total cruciferin transcripts and proteins is observed (Table 3), thought the spatial-temporal distribution of individual cruciferin isoforms have yet to be explored. In food processing, cruciferin functions as a good gelling agent [128] and has the ability to form stronger gels than napin [129], enabling it to be used in a wide array of food products [33]. Cruciferin is also able to improve foaming stability in oil-containing mixtures [130].

**Table 4.** Five manually curated entries for *Brassica napus* L. cruciferin are listed in the Uniprot database (Source: https://www.uniprot.org, accessed September 2021). Length is presented as amino acid count.

| Entry | Entry Name | Protein Names | Length |
|---|---|---|---|
| P33522 | CRU4_BRANA | Cruciferin CRU4 (11S globulin) (12S storage protein) [Cleaved into: Cruciferin CRU4 alpha chain; Cruciferin CRU4 beta chain] | 465 |
| P11090 | CRUA_BRANA | Cruciferin (11S globulin) (12S storage protein) [Cleaved into: Cruciferin subunit alpha; Cruciferin subunit beta] | 488 |
| P33524 | CRU2_BRANA | Cruciferin BnC2 (11S globulin) (12S storage protein) [Cleaved into: Cruciferin BnC2 subunit alpha; Cruciferin BnC2 subunit beta] | 496 |
| P33525 | CRU3_BRANA | Cruciferin CRU1 (11S globulin) (12S storage protein) [Cleaved into: Cruciferin CRU1 alpha chain; Cruciferin CRU1 beta chain] | 509 |
| P33523 | CRU1_BRANA | Cruciferin BnC1 (11S globulin) (12S storage protein) [Cleaved into: Cruciferin BnC1 subunit alpha; Cruciferin BnC1 subunit beta] | 490 |

### 8.2. Napin of B. napus

The second most abundant SSP in canola seed is napin, which accounts for approximately 20% of the seed protein pool [98,131]. Napin is classified as an albumin-type protein reflecting its solubility in water [115]. Low-molecular-weight proteins were initially isolated from canola by Lönnerdal and Janson [132]. Characterization of these strongly basic proteins found them to be composed of two polypeptides, approximately 90 and 30 amino acids in length, respectively, linked by disulfide bridges with a total molecular mass of 12–14 kDa [132]. Subsequent experiments demonstrated that both napin chains were generated from the cleavage of a common precursor polypeptide [133].

Seed storage albumins are encoded by multigene families [134,135]. In canola, a minimum of 10 to 16 genes were initially estimated by Southern blotting to encode napin [136,137] and no updated estimates have been published despite the availability of reference genomes. Seven manually annotated entries for *B. napus* napin are currently listed in the UniProt Knowledgebase (Table 5) (https://www.uniprot.org, accessed on November 2019).

**Table 5.** Seven manually annotated entries for *Brassica napus* L. napin are listed in the Uniprot database (Source: https://www.uniprot.org, accessed September 2021). Length is presented as amino acid length.

| Entry | Entry Name | Protein Names | Length |
|---|---|---|---|
| P24565 | 2SSI_BRANA | Napin-1A (Napin BnIa) [Cleaved into: Napin-1A small chain; Napin-1A large chain] | 110 |
| P09893 | 2SSE_BRANA | Napin embryo-specific (1.7S seed storage protein) [Cleaved into: Napin embryo-specific small chain; Napin embryo-specific large chain] | 186 |
| P17333 | 2SS4_BRANA | Napin (1.7S seed storage protein) [Cleaved into: Napin small chain; Napin large chain] | 180 |
| P27740 | 2SSB_BRANA | Napin-B (1.7S seed storage protein) [Cleaved into: Napin-B small chain; Napin-B large chain] | 178 |
| P01090 | 2SS2_BRANA | Napin-2 (1.7S seed storage protein) [Cleaved into: Napin-2 small chain; Napin-2 large chain] | 178 |
| P01091 | 2SS1_BRANA | Napin-1 (1.7S seed storage protein) [Cleaved into: Napin-1 small chain; Napin-1 large chain] (Fragment) | 133 |
| P80208 | 2SS3_BRANA | Napin-3 (1.7S seed storage protein) (Napin BnIII) (Napin nIII) [Cleaved into: Napin-3 small chain; Napin-3 large chain] | 125 |

The biosynthesis of napin mirrors that of cruciferin [126] but is initiated before the latter (Table 3) and was reviewed in detail by Mylne et al. [134]. Translation of the napin-encoding genes at the ribosomes on the rough endoplasmic reticulum (rER) generates a single pre-proalbumin polypeptide consisting of a signal peptide and the two mature napin chains with linker peptides between each element [4–7]. The signal peptide is removed during translation and the resulting proalbumin is directed into the rER lumen [134]. Inside the lumen, a pattern of eight cysteine residues conserved across similar plant storage albumins allow four disulfide bonds to be formed: two bonds are formed within the large subunit, and two intermolecular bonds are formed between the large and small subunits [134,138]. The polypeptide is then folded with the aid of chaperone proteins and can be subject to post-translational modifications [134]. Subsequently, the proalbumin is transported to the protein storage vacuole where the linker peptides are removed by proteases to form the mature protein [131,134]. The tertiary and quaternary structures of mature napin contribute to its structural stability and resistance to digestion [131,134].

Napin is water soluble, thus enabling it to be incorporated into many food products [130]. In addition, napin also has exceptional foaming capacity and good emulsifying properties [130] though these properties vary depending on the extraction process [139]. These properties enable napin to be used to partially replace more-costly milk proteins [140] and egg white [33] in food processing.

## 9. Genetic Control of Seed Storage Proteins in Canola

Given the potential value of cruciferin and napin in food processing [33,64], the development of cultivars with improved accumulation of either protein can add value to the crop. In order to facilitate breeding for improved seed storage protein in canola, knowledge of the existing genetic variation of the trait is required, in addition to an understanding of the genetic and non-genetic determinates that affect their synthesis and accumulation. To date, a single study examining genetic variation in seed storage protein was conducted in winter rapeseed [141], while similar studies have yet to be undertaken in spring rapeseed germplasm. Seed storage protein content within the seed is dependent on not only the expression of SSP-encoding genes but also on the post-translational processes that generate the quaternary structure from precursor polypeptides, and their subsequent transfer of assembled SSP into storage vacuoles [134,142,143]. The quantitative nature of SSP content suggests complex genetic regulation of the trait and indeed many genes are involved in the synthesis and processing of SSP.

Quantitative trait loci (QTL) are regions within the genome whose occurrences are associated with specific quantitative phenotypes; multiple QTL each contribute some variation to the phenotype, and multiple genes may reside within each locus [144]. The successful identification of QTL associated with important seed quality traits in canola such as oil content [145–149], fatty acid composition [150,151], and glucosinolate levels [146,152–155] have been reported by numerous groups. In contrast, comparatively few reports have focused on the identification of QTL associated with total seed protein in canola [145,148,149,155]. Even fewer are reports of QTL associated with protein quality traits such as SSP composition [155] and amino acid content [156].

Conserved motifs exist in the promoter regions of seed storage protein genes across diverse plant species [157] suggesting that the regulation of seed storage protein biosynthesis is governed by transcriptional mechanisms that evolved early in the evolution of the plant kingdom. The disruption of some of these conserved motifs in the canola cruciferin-encoding *cru1* gene [158] and the napin-encoding *napA* gene [159] led to reduced accumulation of the respective protein product. Multiple transcription factors are known to regulate the expression of seed storage protein genes in *Arabidopsis*: *ABSCISIC ACID INSENSITIVE 3 (ABI3)* [160–163], *FUSCA 3 (FUS3)* [161,164], *LEAFY COTYLEDON 1 (LEC1)* [164–166], *LEC2* [161], and multiple *MYC* transcription factors [167]. A detailed summary on the transcriptional regulation of seed storage proteins was reviewed by Verdier and Thompson [168].

## 10. Non-Genetic Control of Seed Storage Proteins in Canola

Plant growth regulators are capable of modifying transcriptional activity and consequently, seed storage protein accumulation may in part be influenced by plant hormones [168]. Early works on seed storage proteins in canola implicated abscisic acid in the direct regulation of cruciferin and napin [100,169,170]. In addition to its role in inducing the accumulation of seed storage proteins, abscisic acid is also associated with abiotic stress tolerance in plants [169]. Thus, seed storage protein accumulation is influenced by the environment in which the plant is grown, as reported in canola [170] and winter wheat [171]. Indeed, from a breeding perspective, seed storage protein content is a quantitative trait which suggests that the phenotype is influenced by the environment. Research in soybean has demonstrated that differences in soil fertility were capable of increasing total seed protein content [172] as well as altering storage protein composition [135,173]. In canola, crop production on sulfur-limited land not only results in a reduction in glucosinolates and sulfur-containing amino acids [174–176] but can also alter seed storage protein profiles in the crop [177]. Recent work has also correlated nitrogen supply with the expression of seed storage protein genes in canola [114]. Though the extent to which environmental factors influence seed storage protein accumulation has yet to be empirically determined, the aforementioned studies suggest that agronomic practices can be exploited to alter seed protein content and quality.

## 11. Canola Protein as a Novel Food Product

Currently, canola protein is primarily used for animal feed but has the potential to be directly consumed in food products [64,178]. Canola protein does not have a history of human consumption, making it a novel food in the context of most food safety regulations. The procedures and standards for assessing the safety of novel foods, and the authorization required for their marketing, vary by country. Toxicology studies on the safety of a novel food are based on animal feeding experiments. While safety concerns were raised in early animal studies using rapeseed meal (non-canola quality), recent studies using canola-quality meal products concluded the ingredient to be safe. Collectively, the literature suggests that the safety of rapeseed meal was compromised by anti-nutritive compounds and the strict use of canola-quality *B. napus* alleviates safety concerns. Specific safety studies on a cruciferin-rich [179] and a napin-rich [180] canola protein isolate, respectively, found no adverse effects in rats fed with a diet supplemented with up to 20% of either product. Due to the high costs associated with acquiring regulatory approval, authorization of novel foods is typically only sought for countries with potential markets. Canola protein is currently a fledgling in the plant protein sector whose commercialization has been spearheaded by three companies; the history of canola protein commercialization and its challenges have been recently analyzed by Mupondwa et al. [64].

### 11.1. Allergenicity of Canola Seed Storage Proteins

The ability of napin to resist digestion suggests its potential as a food allergen [181,182]. Allergens are foreign proteins that are capable of evoking an adverse response from the immune system after they enter the body [183,184]. Generally, an immediate allergic response happens when an allergen has been encountered twice. The first time an allergen is encountered, the body produces a type of antibody that can bind the allergen called immunoglobulin E (IgE); this first encounter and production of IgE is collectively called sensitization [184]. Following sensitization, if the allergen is encountered a subsequent time, an immune response occurs in which symptoms typically associated with an allergic response, such as urticaria (hives) and rhinitis (runny nose), are experienced [183,184]. Repeated exposure to an allergen can result in increased severity of the symptoms. The molecular mechanisms behind how IgE function during allergies have been reviewed by Gould and Sutton [185].

Napin from canola has been identified as a potential allergen [128] based on its ability to bind IgE from sensitized patients [186]. Further, napin, in addition to cruciferin, have both been identified as allergenic proteins in cold-pressed canola oil that reacted with IgE from sensitized children [187]. These results were considered by the regulatory bodies when determining the safety of canola protein products, but were not deemed to be a substantial risk [188].

An allergenic reaction can be caused by a protein, to which a patient has not been previously sensitized, if it is sufficiently similar in structure to a known allergen [189,190]. This cross-reactivity suggests that patients who are allergic to other plant globulins and plant albumins, respectively, may react to cruciferin and napin and vice versa [182,190,191]. Mustard was added to a list of priority allergens in Canada [192] following a systematic review completed by Health Canada [193], who found sufficient scientific evidence of its allergenicity to be relevant to the Canadian public. The 2S albumin of mustard seed is the major allergen of mustard [191,194] and is similar in its amino acid sequence to napin [110,182], suggesting that people who are sensitized to mustard may also be allergic to napin.

Although the allergenicity of proteins in canola oil have been demonstrated and concerns of their potential to cross-react in patients sensitized to related protein allergens have been raised, it is noteworthy to consider that there have been no reports on allergies to canola oil, likely due to the refining process [187]. Furthermore, the long history of mustard oil consumption in certain parts of the world and the nutritional value of canola oil outweigh concerns of its allergenicity [195]; however, questions regarding the safety of concentrated canola protein products need to be evaluated, given their novel nature.

### 11.2. Regulation of Novel Foods in the United States

Canola meal and canola proteins have not had an extensive history of consumption and are considered novel foods by the United States Food and Drug Administration (FDA). Novel foods in the United States that are deemed generally recognized as safe (GRAS) by the FDA fall outside the purview of the Federal Food, Drug, and Cosmetic Act [196]. For a novel food to be granted GRAS status, its safety must be demonstrated empirically and be recognized as safe by the scientific community. Under the current GRAS Notification program, the FDA does not conduct scientific testing on the novel food to determine its safety, but rather, the Agency reviews the safety data it receives [196]. The general process required for an ingredient to acquire GRAS status involves three steps: first, the sponsor of the product performs or submits relevant scientific studies on the safety of the novel food in the form and dose it is intended to be used; second, the data are submitted as part of a GRAS notice to the FDA for review; and third, the FDA will respond with a letter stating whether or not it has questions regarding the GRAS conclusion of the novel food based on the submitted notice [196]. A letter indicating that the agency has no questions equates to acknowledgement and acceptance of the GRAS status of the ingredient. To date, six GRAS notices regarding *B. napus* products, of which three pertain to meal protein, and the FDA's response to each one is found in the FDA's online GRAS Notice Inventory (Table 6) (https://www.fda.gov/food/generally-recognized-safe-gras/gras-notice-inventory, accessed on 1 June 2021).

### 11.3. Regulation of Novel Food in the European Union

As in the United States, novel foods must be deemed safe before they can be used in the European Union. Canola protein and canola protein isolates are considered novel foods in the EU, which by definition are foods that are not "used for consumption to a significant degree" prior to 15 May 1997 (Regulation (EC) No 258/97) [197]. The safety determination process in the EU is generally similar to that of the GRAS Notification Program: sponsors of novel food submit data on the safety of the ingredient under its intended use to the European Commission, who then forwards the application to the European Food Safety Authority (EFSA). The EFSA assesses the data and publishes its finding as a Scientific Opinion; a committee votes whether to accept the finding; and the Commission issues authorization based on the vote. To date, the EFSA has published only one Scientific

Opinion on a protein isolate derived from a mix of *B. napus* and *B. rapa*, finding the product to be safe under its intended use [188].

**Table 6.** Six canola (*Brassica napus* L.) products are listed in the United States Food and Drug Administration's (FDA) Generally Recognized as Safe (GRAS) notice inventory since the inventory was established in 1998. In all applications, the FDA had no questions regarding the GRAS status of the product. Current as of September 2021.

| GRAS Number | Notifier | Substance | Intended Use | Date of Closure | FDA Response |
|---|---|---|---|---|---|
| 683 | DSM Innovation Company | Canola protein isolate | Dietary protein Food processing | 2017 | no questions |
| 682 | Cargill Inc. | Lecithin from canola | Food processing Dietary fat Dietary choline | 2017 | no questions |
| 533 | American Lecithin Company | Lecithin from canola | Food processing Dietary fat Dietary choline | 2015 | no questions |
| 425 | Danone Trading B.V. | Canola oil (low-erucic-acid rapeseed oil) | Dietary fat (infant formula) | 2012 | no questions |
| 386 | BioExx Specialty Proteins, Ltd. | Canola protein isolate and hydrolyzed canola protein isolate | Food processing Dietary protein | 2011 | no questions |
| 327 | Archer Daniels Midland Company | Cruciferin-rich canola/rapeseed protein isolate and napin-rich canola/rapeseed protein isolate | Dietary protein | 2010 | no questions |

*11.4. Regulation of Novel Food in Canada*

In Canada, Health Canada is the regulatory body overseeing the safety of food additives and its decisions are enforced by the Canadian Food Inspection Agency (CFIA). Purified canola protein and canola protein isolates are considered novel foods in Canada as they do not have a history of use for human consumption [198]. The process to request approval for novel foods approximates that of the United States and European Union: first, the applicant submits pertinent scientific studies and data on the safety of the ingredient to the Food Directorate at Health Canada; second, the application is forwarded for scientific review after the Food Directorate has verified the information is complete; and third, the results of the review are sent back to the Food Directorate, who then decides on the authorization of the ingredient in question. Similar to the GRAS Notice Program, the decision of the Food Directorate is communicated as a Letter of No Objection for approved novel foods. An inventory of novel food decisions that received no objection (https://www.canada.ca/en/health-canada/services/food-nutrition/genetically-modified-foods-other-novel-foods/approved-products.html, accessed on 1 June 2021) is available online. Currently, no canola protein products have been approved as a novel food in Canada.

## 12. Canola Meal as a Protein Source in Animal Husbandry

Protein is an integral part of animal diets; however, it is typically a costly component of feed [199] and efforts to explore low-cost protein supplements have been untaken. The low cost and abundance of rapeseed and canola meal render it an economical protein source [33,64,200]. Mammals that are commercially raised for food production can largely be classified based on their ability to acquire nutrients from plant material: ruminants, such as cattle, have the capacity to ferment plant material prior to digestion allowing for successful nutrient uptake, while non-ruminants, such as swine, lack this ability [22]. This difference in digestive anatomy has implications in the use of canola meal as a protein supplement in the respective feeds of these animals: specifically, differing sensitivities to glucosinolates between ruminants and non-ruminants generally dictate the relative quantity of canola meal that can be incorporated into feed [51,201]. In addition to cattle and swine, poultry and fish are also raised for food on commercial scales and the feasibility of using canola meal in these production systems has also been studied.

### 12.1. Cattle

The United States is the largest importer of Canadian canola meal (Table 980-0012, Statistics Canada, 2019). The use of canola meal as a protein supplement has been widely adopted by the cattle (*Bos taurus* Linnaeus, 1758) industry since the development of canola [201]. Specifically, canola meal is one of the standard protein supplements for dairy cattle production [202]. Given the ruminant nature of cattle, the incorporation of plant-based protein supplements such as soybean meal is a well-established practice. A meta-analysis conducted on dairy cattle found that the use of canola meal was superior to soybean meal as a protein supplement as measured by feed intake, milk yield, and milk protein yield [203]. In cattle produced for meat, the substitution of barley as the protein source with canola meal during the growing period did not increase feed efficacy compared to the use of barley [204]. Although the conversion of feed to animal protein was not improved, the use of canola meal in feedlot cattle production may still be a more economical and sustainable alternative than using barley. Taken together, these studies support the continued use of canola meal as a protein supplement in cattle production.

### 12.2. Swine

Soybean meal is widely used as an economical protein source in swine (*Sus scrofa domesticus* Erxleben, 1777) production and remains a reference for novel plant-based protein supplements. Canola meal can partially replace soybean meal in the swine diet without ill-effect on animal health and production efficiency [75]; however complete replacement is hampered by its high fibre content and the presence of various secondary metabolites, especially glucosinolates based on feeding experiments [205]. Swine are more sensitive to glucosinolates compared to cattle [51,205] and residual glucosinolates in canola meal-containing feed will drive the animals to choose soybean meal-supplemented feed when presented with the option [206]. Additional processing steps during meal production such as toasting were able to mitigate feed rejection presumably due to the decomposition of residual glucosinolates; however, production efficacy did not supersede that of soybean meal [75]. Technical advances in canola meal processing will enable it to be a competitive and cost-efficient alternative to soybean meal [205]

### 12.3. Poultry

Poultry (*Gallus gallus domesticus* Linnaeus, 1758) ranks second as the most consumed animal protein globally [207]. Protein in the chicken diet is often supplied from a mix of animal and plant sources [199,208], the latter of which is primarily from soybean due to its nutritional properties [209]. Although chickens grow faster with some animal protein in their diet, work has been conducted to observe the effects of using solely vegetable-sourced protein for chicken [77,208,210–212]. Canola meal has been explored as a plant-based protein source for chicken production and was found to be generally comparable to soybean meal [212,213]. An examination of the lower digestive tract of chickens fed canola meal failed to identify intact proteins typical of canola seed, suggesting that the meal protein was completely digested by the animal [211]. Of cattle, swine, and chicken, the growth of the latter appears to be least affected by the use of canola meal as the sole source of feed protein, as evidenced by the possibility of using canola meal as the sole protein source in feed [83].

### 12.4. Aquaculture

Fish currently account for approximately 20% of the animal protein consumed globally and its production is split approximately in half between aquaculture (farmed fish) and capture fisheries (wild caught fish) [214]. The standard protein source for aquaculture has historically been fish meal; however, high costs, limited supply, and sustainability concerns led to a decline in the use of fish meal in favour of plant protein sources and other novel feedstuff [199,214,215]. The complete replacement of fish meal with other protein sources has been reported to be successful across various aquaculture species, particularly those of lower trophic levels [199]. As with land animals, fish are sensitive to glucosinolates and high levels in their feed from the inclusion of canola meal can negatively impact production efficiency [51]. Nonetheless, canola meal and different canola protein isolates have proven successful as a protein supplement in various aquaculture species [215], including rainbow trout (*Oncorhynchus mykiss* Walbaum. 1792) [216–218], and high-value crustaceans such as mitten crab (*Eriocheir sinensis* H. Milne-Edwards, 1853) [219] and shrimp species (*Litopenaeus stylirostris* Stimpson, 1871; *L. vannamei* Boone, 1931) [220,221].

## 13. Seed Storage Proteins in Food Processing

In addition to fulfilling its role as a macronutrient, protein can also be supplemented in food products to improve its nutritional profile. Plant-based protein isolates, namely derived from soybean and pea and comprised mostly of SSP, are ubiquitously available on the health food market in either its pure form or supplemented in food products [222]. The continued availability of these products on supermarket shelves speaks to the palatability of plant-based proteins and their acceptance by the general consumer.

Proteins also serve a non-nutritional role in food by directly controlling the physical and functional l food properties [64,198]. To ensure food products, particularly meat analogues, meet the textural expectations of consumers [223], commercial processors use different proteins [222] in addition to different mechanical processes [224]. As different SSP have different physiochemical properties based on their amino acid composition and consequently their structure, not all SSP can fulfill the same functional role [34,225]. More importantly, plant-based protein isolates contain a mixture of SSP whose constituents may be mutually antagonistic in their functionalities and raw isolates may therefore be unfit for use as a structural additive in commercial food processing despite its nutritive contribution [225]. One solution towards the adoption of plant protein for food processing applications is to supplement single SSP, or single SSP-enriched isolates rather than raw isolates. In this way, control over the texture of the final food product can be better maintained. Furthermore, SSP can be strategically used to improve not only the absolute protein content of the food product, but also the amino acid profile of the product [226,227], allowing for marketing to specific consumer bases.

## 14. Separation and Quantification of Seed Storage Proteins

Globulins and albumins can be separated based on their difference in solubility in dilute saline solution [115]. Indeed, napin and cruciferin have been successfully separated on the basis of solubility in salt [34,39,228–230]. Separation of cruciferin and napin can also be facilitated by the differences in solubility in different pH [73,96,228–230]. This appears to be the basis, at least partially, of patents granted for the commercial separation of these major canola SSP. A comprehensive review on separation technologies specific to canola protein and a selected list of relevant patents was recently compiled by Mupondwa et al. [64].

On a bench scale, the separation of cruciferin and napin can be accomplished by means of chromatography [98]. Although chromatographic separation and purification results in a highly purified single-SSP product, the cost of such methods, the technical expertise required for their operation, and the low recovery rate [96] limit such methods from being used on a commercial scale until broader market demand for these proteins develop. The ease of protein separation can be improved if the native SSP pool has reduced species complexity or consists of a single protein species.

Breeding efforts towards altering SSP composition require a method for quantifying individual canola proteins. Most breeding programs lack the expertise and infrastructure required to perform chromatography, and although accurate, such analytical methods lack the throughput required for phenotyping large populations. In current breeding programs, SSP quantification is performed by SDS-PAGE followed by Coomassie staining and densitometry [155]. However, this method fails to distinguish between different proteins with similar molecular masses and preferential binding of the dye to aromatic amino acids can skew the accuracy of the assay [231]. Immunological methods to quantify individual SSP may offer the accuracy and throughput needed for breeding programs.

## 15. Immunodetection and Quantification of Seed Storage Proteins

Electrophoretic separation of proteins followed by immunodetection of target antigens was first described by Towbin et al. [232] to analyze ribosomal proteins. To date, this variant of Western blotting remains a prevalent technique in the literature being employed in approximately 10% of all protein research articles [233]. Despite criticisms of its reproducibility due to variation in antibody quality [233], Western blotting can be a valuable technique for the relative quantification of proteins in mixtures provided proper extraction protocols, standard curves, and normalization protocols are implemented [234]. Specifically, the use of Western blotting to quantify individual SSP within total seed protein mirrors conventional size exclusion chromatography (SEC) methods [141] in that both techniques rely on the separation of proteins by molecular mass prior to estimating the abundance of target proteins of a given theoretical mass. Arguably, Western blotting requires less infrastructure to perform and the use of an antibody allows for greater discrimination of proteins compared to mass alone. Conversely, Western blotting has been implemented to qualitatively elucidate basic knowledge of SSP structure [127,235,236] and post-translational modifications [237]. Similarly, immunological methods have also been employed to study the spatial distribution of SSP [122] and quantify temporal patterns in their accumulation [98,238,239] through seed development.

## 16. Manipulation of Seed Storage Proteins in Select Crop Species

The challenges of chemical separation of SSP can be circumvented altogether with plants whose seed protein pool contains a single protein species. Seed storage proteins are regulated on a genetic level, and therefore genetic technologies can be employed to alter SSP composition in plants. Efforts were undertaken to genetically alter the SSP profile of various crop species, for functional [240,241] or nutritive purposes [240,242,243] have been successful.

### 16.1. Seed Storage Protein Manipulation through Conventional Breeding

Conventional breeding relies on cycles of crossing and selection to generate distinct genotypes with improvements in desired traits and genetic variability is required in plant breeding to facilitate genetic gain in the trait of interest [244]. To date, only one survey of SSP variability in rapeseed has been conducted in France [141], which found historic rapeseed cultivars to be relatively richer in napin content compared to modern canola-quality cultivars. Currently, the genetic variation that exists for SSP in Canadian germplasm remains unknown.

Diversity in SSP composition has been reported in various crop species such as wheat [245] and soybean [246], implying similar diversity may exist in canola. Wild soybean accessions with unique SSP profiles have been successfully incorporated into breeding programs to generate segregating populations with variable seed protein profiles [247] indicating that the manipulation of storage protein composition can be achieved by conventional breeding.

Intentional selection for SSP composition in canola has never been reported; however, selection for oil content is believed to have also inadvertently selected for increased cruciferin content [141]. A transcriptomic study of the breeding response in Chinese winter rapeseed over two decades of selection for improved fatty acid profiles and oil content revealed minimal changes to the transcription of cruciferin and napin, though actual levels of SSP were not determined [248]. Interestingly, protein and oil content both increased concomitantly with overall yield [248], suggesting that the inverse relationship between the two traits can be broken. Furthermore, cruciferin and napin levels were found to be highly heritable, suggesting the possibility for the successful genetic improvement of the levels of either protein through conventional breeding [155].

### 16.2. Soybean

Soybean is primarily grown as an oilseed in North America and as an important protein source in Asia [249]. Approximately 70% of the soybean SSP pool is composed collectively of the multimeric globulins glycinin (11S) and β-conglycinin (7S) [250,251]. While soy protein is versatile in both its functional and edible properties [252], a significant portion is directed towards tofu production. The marketability and consumer acceptance of tofu is mostly based on its texture; thus, research has focused on the relationship between soybean protein and its functional properties in tofu production. Seed storage protein composition is known to play a critical role in controlling the texture of tofu [253], and more recent work has demonstrated that the subunit composition of individual SSP also plays a critical role [254]. An effort to change the SSP profile of soybeans through a combination of mutagenesis and convention breeding has led to the development of genotypes with improved tofu-making qualities [247,254–256].

Simultaneously with improving protein functionality, efforts to improve the nutritional value of soybean protein have also been undertaken. The improvement of lysine content in the total seed protein of soybean was made possible by altering its feedback regulatory mechanism during biosynthesis [227]. Effort aimed at improving the methionine content of soybean protein was made through the expression of a chimeric gene encoding a methionine-rich δ-zein SSP from corn under the control of an endogenous soybean β-conglycinin promoter [257]. Increased accumulation of foreign proteins was enabled by the successful suppression of β-conglycinin production by RNAi [258,259]. These works demonstrate the possibility for the continued improvement of soy protein nutrition by enabling the improved accumulation of higher-nutrient or therapeutic proteins [259] at the expense of lower-nutrient endogenous SSP.

### 16.3. Wheat

In wheat, the seed protein pool, colloquially referred to as gluten, is comprised of the SSP gliadins and glutelins [109,260]. Wheat SSP primarily serves a structural rather than nutritive role in food and as such, breeding efforts have centered around improving its functionality. Despite the prevalence of wheat in the diet, the protein constituents of gluten elicits immunogenic reactions in susceptible patients and recent work has focused on the elimination of its reactivity [261,262]. Efforts to improve the functional properties of wheat have been made by various groups through the overexpression of glutenins [263–265]. The ectopic expression of genes encoding high-molecular-weight glutenin subunits resulted in an increased accumulation of the protein [264,265] as well as increased dough elasticity [263] in a dose-dependent manner. The downregulation of gliadins using RNAi has also been shown to improve the functionality of flour [266,267], demonstrating the direct relationship between SSP composition and protein functionality. Furthermore, flour from gliadin-deficient wheat genotypes was demonstrated to have reduced immunogenicity [261,262], suggesting that the manipulation of SSP can be a strategy to generate food products to cater to specialized dietary needs. Collectively, these studies in soybean and wheat provide empirical evidence on the possibility of manipulating seed storage proteins in canola.

### 16.4. Canola

To date, canola remains primarily an oilseed crop and its meal protein largely remains a byproduct [268]. Meal protein has only been considered for use as a nutritional supplement, and consequently efforts to alter the SSP profile of canola has only been examined from a nutritional standpoint. To improve the methionine content of canola meal protein, a chimeric gene containing the coding sequence of a methionine-rich 2S SSP from Brazil nut (*Bertholletia excels* Humb. and Bonpl.) under the control of either a phaseolin promoter [269] or soybean lectin promoter [270] (both promoters whose native functions drives SSP expression in the seed), was transformed into canola. Seeds from the resultant transgenic plants showed elevated levels of methionine [269,270], suggesting that the amino acid profile of canola meal protein was amenable to change and could be altered through the expression of foreign proteins in the seed. Efforts to improve the lysine content of canola meal protein were made by disrupting the feedback-regulation of lysine during biosynthesis, which effectively doubled the total seed lysine content [227]. This indicates that improvements to the levels of individual amino acids can be achieved. More importantly, these works suggest that the plant is able to accumulate higher than normal levels of individual amino acids without severe physiological consequences. Subsequent efforts to improve the quality of canola meal protein were made using advancements in antisense technology with the goal of selectively attenuating individual SSP [226,271]. By suppressing the accumulation of napin transcripts, an increase in cruciferin protein was observed without major effects on fatty acid composition [271]; similarly, reducing the level of cruciferin transcripts resulted in an increase in napin protein content as well as improved levels of methionine, lysine, and cysteine [226]. In both cases, modifications to the SSP did not result in changes to the total macromolecular profile of the seed. Interestingly, co-suppression of both cruciferin and napin together did not lead to a compensatory increase in oil [272], suggesting that the inverse relationship between oil and protein content is not a simple competition for metabolic intermediates.

## 17. Plant Breeding in the Omics Era

Advancements in omics technologies have enabled breeders to study the underlying mechanisms that govern desirable phenotypes on a genomic, transcriptomic, and proteomic level. These technologies improve the efficiency of breeding programs and improve the speed in which new cultivars can be generated.

### 17.1. Genomics

A major goal of genomics in crop agriculture is to correlate genotypic information with phenotypic data [273]. In this way, selections can be made early in the growth cycle on the basis of molecular markers that are impervious to environmental variation, improving both efficiency and accuracy. Early genotyping efforts relied on restriction digestion and hybridization to discern allelic variation. The advent of PCR subsequently enabled the development of amplification-based genotyping platforms with improved throughput and efficiency [273]. Next, genotyping by single-nucleotide polymorphism (SNP) markers was developed as the markers were numerous and widely distributed across plant genomes [274]. Multiplex SNP genotyping using crop-specific bead chip arrays such as the *Brassica napus* 60K Illumina Infinium™ SNP array [275,276] further enhanced the efficacy of acquiring genotypic information in crop plants [277] by enabling thousands of SNPs to be surveyed simultaneously. As next-generation sequencing becomes increasingly more accessible, genotyping with SNP markers may potentially be replaced by whole genome sequence data. While early whole-genome sequencing protocols such as restriction-site associated DNA sequencing [278] and genotyping-by-sequencing [279] used restriction enzymes to reduce the complexity of the genome and cost, rapid improvements in sequencing technologies now enable whole genomes to be economically sequenced at unprecedented speeds [280].

### 17.1.1. Linkage Mapping

Linkage mapping identifies significant correlations between molecular markers and traits of interest in a structured population [281,282]; such populations are typically generated by crossing parents with divergent phenotypes and subsequently fixing recombination events in the segregating progeny through doubled-haploid production or repeated cycles of selfing. The size and relatedness between individuals in the population directly affect the resolution of the mapping study [283]. The mapping population along with the parental genotypes are phenotyped in replicated experiments and genotyped by SNP-chip arrays [284]. A linkage map displaying the physical order of the SNP markers and the genetic distance between them can then be built with various software packages by inferring recombination ratios [285,286]. Correlations between markers and phenotypic data are performed using various statistical methods [286]. Linkage mapping has been successfully employed in canola to identify molecular markers that co-segregate with total seed protein [145–149], seed storage protein content [155], and non-essential amino acid content [156].

17.1.2. Association Mapping

Similar to linkage mapping, association mapping also aims to identify significant correlations between genotypic information and phenotypes [274,281]. However, unlike linkage mapping, which relies on recombination events occurring in a structure bi-parental population, association mapping (also referred to as genome-wide association studies; GWAS) takes into consideration all historic recombination events in a population of genetically diverse individuals [287]. The practical implications of this difference is twofold: first, by using a population of genetically diverse individuals, no time investment to generate biparental mapping populations is needed, which effectively shortens the time required for association mapping; second, by using genetically diverse genotypes and considering all historical recombination events, the resolution of association mapping is higher than that of linkage mapping, and causal SNPs can be identified rather than loci [281,287,288].

Genome-wide association studies were initially proposed for use in human genetics in the 1990s [289]. At a similar time, the first association studies were reported on grain crops [287]. Advancement in statistical methods to account for genetic challenges that are unique to crops were first incorporated into a GWAS for flowering time in maize a decade later [290]. To date, GWAS has been successfully and regularly implemented to study many major crop species [282,291,292].

The population size required for GWAS varies depending on the goal of the study and the inherent genetic structure of the population: 300 individuals are sufficient for candidate gene validation and 1000–5000 individuals are recommended for marker discovery [293]. Although larger populations are able to improve the power to detect QTL in canola by GWAS [294], populations of 200–500 individuals have been typical in recent studies (Table 7). Phenotypic data are collected from replicated experiments and genotypic data, often in the form of SNP markers, are typically acquired through high-throughput SNP-chip arrays that are crop specific [284]. Recent studies have used whole genome sequencing and transcriptome sequencing in place of SNP-chip arrays to generate genotypic data [295]. The significance of the association between each SNP and the phenotype is then tested using different statistical models while considering the relatedness between each individual of the population and possible cryptic population structures [274,282,287]. Finally, the results of GWAS are typically visualized in a Manhattan plot where the significance of the marker association is plotted against the physical position of the marker on the genome, thus offering a view of all the tested markers across the genome [296].

In canola, GWAS has been successfully implemented to identify molecular markers and loci associated with a variety of agronomically-important traits, and seed quality traits (Table 7); however, studies focusing on seed storage protein-related traits are lacking. In other major crops such as rice [297] and legumes [143], GWAS identified markers that underlie variation in storage protein accumulation. Furthermore, markers associated with amino acid content were identified using GWAS in maize [298], wheat [299], and soybean [300], as well as *Arabidopsis* [301]. The lack of similar association studies in canola focused on seed storage protein traits represents a void in the literature that warrants investigation.

**Table 7.** Genome-wide association studies have been successfully implemented in *Brassica napus* L. to identify single-nucleotide polymorphism (SNP) markers associated with seed quality traits. Studies listed in this table were performed using populations of various sizes and genotypic data acquired using the Brassica 60K Illumina Infinium™ SNP genotyping array [276].

| Trait | Population Size | Reference |
| --- | --- | --- |
| Erucic acid content | 215 | [302] |
| Erucic acid content | 203 | [303] |
| Erucic acid content | 472 | [304] |
| Fatty acid composition | 435 | [305] |
| Fatty acid composition | 520 | [59] |
| Fatty acid content | 370 | [306] |
| Fibre content | 520 | [307] |
| Glucosinolate | 521 | [308] |
| Glucosinolate | 203 | [303] |
| Glucosinolate | 520 | [309] |
| Glucosinolate | 1425 | [310] |
| Glucosinolate | 203 | [311] |
| Glucosinolate | 203 | [312] |
| Glucosinolates | 215 | [302] |
| Oil content | 472 | [304] |
| Oil content | 521 | [313] |
| Oil content | 105 | [314] |
| Oil content | 370 | [59] |
| Oil content | 521 | [313] |
| Oil content | 203 | [311] |
| Oil content | 203 | [315] |
| Protein content | 370 | [59] |

## 18. Marking of a New Era of Crop Improvement with Genome Editing

Early efforts in the genetic manipulation of SSP profiles relied primarily on the use of antisense technology [226,240,271,316–319] and RNA interference [240,242,243,316,320,321]. These techniques relied on the degradation of gene transcripts to reduce the accumulation of select SSP. Advancements in genome editing technologies (succinctly reviewed by Langner et al. [322]), namely the Clustered regularly interspaced short palindromic repeats (CRISPR)/CRISPR-associated (Cas) system [323] and its derivatives, now enable the modification of SSP at the genetic level by inducing deleterious mutations in the associated genes. Briefly, the CRISPR/Cas9 system relies on a short single guide RNA (sgRNA) complex to guide the Cas9 endonuclease to the genomic region to which the former is complementary; after a double-stranded break is induced at the target site by Cas9, the host's innate DNA repair mechanism repairs the break in an error-prone fashion, resulting in deleterious mutations that effectively silence the gene [323].

Initially applied to diploid animal and plant systems, CRISPR/Cas9 technology has also proven to be highly effective in targeting multiple homeologous alleles in polyploid *B. napus* [324]. Successful editing of the *ALCATRAZ* gene to improve shattering resistance [325], the *BnWRKY11* and *BnWRKY70* transcription factors to improve *Sclerotinia sclerotiorum* resistance [326], the *Fatty Acid Desaturase 2* gene to modify fatty acid profile [327] and other gene targets [328,329] in *B. napus* have recently been reported. Given the ease of use of the CRISPR/Cas9 system and its numerous successful applications in *B. napus*, genome editing technology has potential to be used as a tool to modify the SSP profile of canola.

Despite the increasing commercial interest in canola protein, breeding efforts towards improving canola meal protein quality are lacking. Research into the development of cultivars with altered SSP profiles will not only facilitate advancements in protein separation technologies, but also encourage the adoption and use of canola protein as a food product.

**Author Contributions:** K.K.Y.S. drafted this work as part of the literature review from his Ph. D. dissertation. R.W.D. revised this work. All authors have read and agreed to the published version of the manuscript.

**Funding:** K.K.Y.S. was supported by an Alexander Graham Bell Canada Graduate Scholarship from the Natural Sciences and Engineering Research Council of Canada. Research funding was received from the Natural Sciences and Engineering Research Council of Canada, Bunge, DL seeds, the Canola Council of Canada and Canola Cluster Funding.

**Data Availability Statement:** All data in this review has been retrieved from previous publications and cited.

**Acknowledgments:** We thank Ashley Ammeter and Jia Sun (University of Manitoba) for their input and discussion during the preparation of this manuscript.

**Conflicts of Interest:** The authors declare no conflict of interest. The sponsors had no role in the design, execution, interpretation, or writing of the study.

# References

1. United Nations, Department of Economic and Social Affairs, Population Division World Population Prospects 2019. Available online: https://population.un.org/wpp/ (accessed on 10 October 2021).
2. Fedoroff, N.V.; Battisti, D.S.; Beachy, R.N.; Cooper, P.J.M.; Fischhoff, D.A.; Hodges, C.N.; Knauf, V.C.; Lobell, D.; Mazur, B.J.; Molden, D.; et al. Radically rethinking agriculture for the 21st century. *Science* **2010**, *327*, 833–834. [CrossRef]
3. Godfray, H.C.J.; Beddington, J.R.; Crute, I.R.; Haddad, L.; Lawrence, D.; Muir, J.F.; Pretty, J.; Robinson, S.; Thomas, S.M.; Toulmin, C. Food security: The challenge of feeding 9 billion people. *Science* **2010**, *327*, 812–818. [CrossRef]
4. Tamburino, L.; Bravo, G.; Clough, Y.; Nicholas, K.A. From population to production: 50 years of scientific literature on how to feed the world. *Glob. Food Sec.* **2020**, *24*. [CrossRef]
5. Gerten, D.; Heck, V.; Jägermeyr, J.; Bodirsky, B.L.; Fetzer, I.; Jalava, M.; Kummu, M.; Lucht, W.; Rockström, J.; Schaphoff, S.; et al. Feeding ten billion people is possible within four terrestrial planetary boundaries. *Nat. Sustain.* **2020**, *3*, 200–208. [CrossRef]
6. Röös, E.; Bajželj, B.; Smith, P.; Patel, M.; Little, D.; Garnett, T. Greedy or needy? Land use and climate impacts of food in 2050 under different livestock futures. *Glob. Environ. Chang.* **2017**, *47*, 1–12. [CrossRef]
7. Springmann, M.; Clark, M.; Mason-D'Croz, D.; Wiebe, K.; Bodirsky, B.L.; Lassaletta, L.; de Vries, W.; Vermeulen, S.J.; Herrero, M.; Carlson, K.M.; et al. Options for keeping the food system within environmental limits. *Nature* **2018**, *562*, 519–525. [CrossRef] [PubMed]
8. Ramankutty, N.; Mehrabi, Z.; Waha, K.; Jarvis, L.; Kremen, C.; Herrero, M.; Rieseberg, L.H. Trends in Global Agricultural Land Use: Implications for Environmental Health and Food Security. *Annu. Rev. Plant Biol.* **2018**, *69*, 789–815. [CrossRef]
9. Fitton, N.; Alexander, P.; Arnell, N.; Bajzelj, B.; Calvin, K.; Doelman, J.; Gerber, J.S.; Havlik, P.; Hasegawa, T.; Herrero, M.; et al. The vulnerabilities of agricultural land and food production to future water scarcity. *Glob. Environ. Chang.* **2019**, *58*, 101944. [CrossRef]
10. Röös, E.; Bajželj, B.; Smith, P.; Patel, M.; Little, D.; Garnett, T. Protein futures for Western Europe: Potential land use and climate impacts in 2050. *Reg. Environ. Chang.* **2017**, *17*, 367–377. [CrossRef]
11. Report of the World Food Conference1975. Available online: https://digitallibrary.un.org/record/701143?ln=en (accessed on 10 October 2021).
12. Food Security Information Network Global report on food crises 2018. Available online: http://www.fao.org/3/a-br323e.pdf (accessed on 10 October 2021).
13. Trade Reforms and Food Security: Conceptualizing the Linkages 2003. Available online: http://www.fao.org/docrep/005/y467 1e/y4671e06.htm. (accessed on 10 October 2021).
14. Semba, R.D. The rise and fall of protein malnutrition in global health. *Ann. Nutr. Metab.* **2016**, *69*, 79–88. [CrossRef]
15. Hwalla, N.; El Labban, S.; Bahn, R.A. Nutrition security is an integral component of food security. *Front. Life Sci.* **2016**, *9*, 167–172. [CrossRef]
16. Crichton, M.; Craven, D.; Mackay, H.; Marx, W.; De Van Der Schueren, M.; Marshall, S. A systematic review, meta-analysis and meta-regression of the prevalence of protein-energy malnutrition: Associations with geographical region and sex. *Age Ageing* **2019**, *48*, 38–48. [CrossRef] [PubMed]
17. Gerbens-Leenes, P.W.; Nonhebel, S.; Krol, M.S. Food consumption patterns and economic growth. Increasing affluence and the use of natural resources. *Appetite* **2010**, *55*, 597–608. [CrossRef]
18. He, Y.; Yang, X.; Xia, J.; Zhao, L.; Yang, Y. Consumption of meat and dairy products in China: A review. *Proc. Nutr. Soc.* **2016**, *75*, 385–391. [CrossRef]

19. Nelson, G.; Bogard, J.; Lividini, K.; Arsenault, J.; Riley, M.; Sulser, T.B.; Mason-D'Croz, D.; Power, B.; Gustafson, D.; Herrero, M.; et al. Income growth and climate change effects on global nutrition security to mid-century. *Nat. Sustain.* **2018**, *1*, 773–781. [CrossRef]

20. Sabaté, J.; Sranacharoenpong, K.; Harwatt, H.; Wien, M.; Soret, S. The environmental cost of protein food choices. *Public Health Nutr.* **2015**, *18*, 2067–2073. [CrossRef] [PubMed]

21. Pimentel, D.; Pimentel, M. Sustainability of meat-based and plant-based diets and the environment. *Am. J. Clin. Nutr.* **2003**, *78*, 660–663. [CrossRef] [PubMed]

22. Wu, G.; Bazer, F.W.; Cross, H.R. Land-based production of animal protein: Impacts, efficiency, and sustainability. *Ann. N. Y. Acad. Sci.* **2014**, *1328*, 18–28. [CrossRef]

23. Sabaté, J.; Soret, S. Sustainability of plant-based diets: Back to the future. *Am. J. Clin. Nutr.* **2014**, *100*, 476–482. [CrossRef] [PubMed]

24. Shepon, A.; Eshel, G.; Noor, E.; Milo, R. The opportunity cost of animal based diets exceeds all food losses. *Proc. Natl. Acad. Sci. USA* **2018**, *115*, 3804–3809. [CrossRef]

25. Graham, F.; Russell, J.; Holdsworth, M.; Menon, M.; Barker, M. Exploring the relationship between environmental impact and nutrient content of sandwiches and beverages available in cafés in a UK university. *Sustainability* **2019**, *11*. [CrossRef]

26. Nelson, M.E.; Hamm, M.W.; Hu, F.B.; Abrams, S.A.; Griffin, T.S. Alignment of healthy dietary patterns and environmental sustainability: A systematic review. *Adv. Nutr.* **2016**, *7*, 1005–1025. [CrossRef] [PubMed]

27. Bentham, J.; Singh, G.M.; Danaei, G.; Green, R.; Lin, J.K.; Stevens, G.A.; Farzadfar, F.; Bennett, J.E.; Di Cesare, M.; Dangour, A.D.; et al. Multidimensional characterization of global food supply from 1961 to 2013. *Nat. Food* **2020**, *1*, 70–75. [CrossRef]

28. David, L.A.; Maurice, C.F.; Carmody, R.N.; Gootenberg, D.B.; Button, J.E.; Wolfe, B.E.; Ling, A.V.; Devlin, A.S.; Varma, Y.; Fischbach, M.A.; et al. Diet rapidly and reproducibly alters the human gut microbiome. *Nature* **2014**, *505*, 559–563. [CrossRef] [PubMed]

29. Food and Agriculture Organization Pulses Nutritious Seeds for a Sustainable Future. Available online: https://www.fao.org/3/i5528e/i5528e.pdf (accessed on 10 October 2021).

30. Blackstone, N.T.; El-Abbadi, N.H.; McCabe, M.S.; Griffin, T.S.; Nelson, M.E. Linking sustainability to the healthy eating patterns of the Dietary Guidelines for Americans: A modelling study. *Lancet Planet. Health* **2018**, *2*, e344–e352. [CrossRef]

31. Tuomisto, H.L. Importance of considering environmental sustainability in dietary guidelines. *Lancet Planet. Health* **2018**, *2*, e331–e332. [CrossRef]

32. Canada's Dietary Guidelines for Health Professionals and Policy Makers 2019. Available online: https://food-guide.canada.ca/en/guidelines/ (accessed on 10 October 2021).

33. Wanasundara, J.P.D.; McIntosh, T.C.; Perera, S.P.; Withana-Gamage, T.S.; Mitra, P. Canola/rapeseed protein-functionality and nutrition. *OCL* **2016**, *23*, D407. [CrossRef]

34. Tan, S.H.; Mailer, R.J.; Blanchard, C.L.; Agboola, S.O. Canola Proteins for Human Consumption: Extraction, Profile, and Functional Properties. *J. Food Sci.* **2011**, *76*. [CrossRef] [PubMed]

35. Book of Abstracts of 15th International Rapeseed Congress 2019. Available online: https://www.openagrar.de/receive/openagrar_mods_00049944 (accessed on 10 October 2021).

36. Pudel, F. Rapeseed proteins recent results on extraction and application. In *Canola and Rapeseed*, 1st ed.; Thiyam-Holländer, U., Eskin, N.A.M., Matthäus, B., Eds.; CRC Press: Boca Raton, FL, USA, 2012; pp. 184–199.

37. Jones, O.G. Recent advances in the functionality of non-animal-sourced proteins contributing to their use in meat analogs. *Curr. Opin. Food Sci.* **2016**, *7*, 7–13. [CrossRef]

38. Chabanon, G.; Chevalot, I.; Framboisier, X.; Chenu, S.; Marc, I. Hydrolysis of rapeseed protein isolates: Kinetics, characterization and functional properties of hydrolysates. *Process Biochem.* **2007**, *42*, 1419–1428. [CrossRef]

39. Yang, C.; Wang, Y.; Vasanthan, T.; Chen, L. Impacts of pH and heating temperature on formation mechanisms and properties of thermally induced canola protein gels. *Food Hydrocoll.* **2014**, *40*, 225–236. [CrossRef]

40. Fetzer, A.; Herfellner, T.; Stäbler, A.; Menner, M.; Eisner, P. Influence of process conditions during aqueous protein extraction upon yield from pre-pressed and cold-pressed rapeseed press cake. *Ind. Crops Prod.* **2018**, *112*, 236–246. [CrossRef]

41. Chalhoub, B.; Denoeud, F.; Liu, S.; Parkin, I.A.P.; Tang, H.; Wang, X.; Chiquet, J.; Belcram, H.; Tong, C.; Samans, B.; et al. Early allopolyploid evolution in the post-Neolithic Brassica napus oilseed genome. *Science* **2014**, *345*, 950–953. [CrossRef] [PubMed]

42. Craig, B.M. Production and utilization of rapeseed in Canada. *J. Am. Oil Chem. Soc.* **1971**, *48*, 737–739. [CrossRef]

43. Nieschlag, H.J.; Wolff, I.A. Industrial uses of high erucic oils. *J. Am. Oil Chem. Soc.* **1971**, *48*, 723–727. [CrossRef]

44. Daun, J.K. Glucosinolate levels in western Canadian rapeseed and canola. *J. Am. Oil Chem. Soc.* **1986**, *63*, 639–643. [CrossRef]

45. Stefansson, B.R.; Hougen, F.W.; Downey, R.K. Note on the isolation of rape plants with seed oil free from erucic acid. *Can. J. Plant Sci.* **1961**, *41*, 218–219. [CrossRef]

46. de Wildt, D.J.; Speijers, G.J.A. Influence of dietary rapeseed oil and erucic acid upon myocardial performance and hemodynamics in rats. *Toxicol. Appl. Pharmacol.* **1984**, *74*, 99–108. [CrossRef]

47. EFSA Panel on Contaminants in the Food Chain Erucic acid in feed and food. *EFSA J.* **2016**, *14*. [CrossRef]

48. Jensen, S.K.; Liu, Y.-G.; Eggum, B.O. The effect of heat treatment on glucosinolates and nutritional value of rapeseed meal in rats. *Anim. Feed Sci. Technol.* **1995**, *53*, 17–28. [CrossRef]

49. Vermorel, M.; Davicco, M.-J.J.; Evrard, J. Valorization of rapeseed meal. 3. Effects of glucosinolate content on food intake, weight gain, liver weight and plasma thyroid hormone levels in growing rats. *Reprod. Nutr. Dev.* **1987**, *27*, 57–66. [CrossRef]

50. Lo, M.; Bell, J.M. Effects of various dietary glucosinolate on growth, feed intake, and thyroid function of rats. *Can. J. Anim. Sci.* **1972**, *52*, 295–302. [CrossRef]

51. Tripathi, M.K.; Mishra, A.S. Glucosinolates in animal nutrition: A review. *Anim. Feed Sci. Technol.* **2007**, *132*, 1–27. [CrossRef]

52. Stefansson, B.R.; Kondra, Z.P. Tower summer rape. *Can. J. Plant Sci.* **1975**, *344*, 343–344. [CrossRef]

53. Rakow, G. *Classical Genetics and Traditional Breeding*; Edwards, D., Batley, J., Parkin, I., Kole, C., Eds.; CRC Press: Boca Raton, FL, USA, 2011; ISBN 9780429067310.

54. Consensus Document on Key Nutrient and Key Toxicants in Low Erucic Acid Rapeseed (Canola) 2001. Available online: https://www.oecd.org/env/ehs/biotrack/46815125.pdf (accessed on 10 October 2021).

55. Food and Agriculture Organization of the United Nations FAOSTAT Database. Available online: http://www.fao.org/faostat/en/#data/QC (accessed on 10 October 2021).

56. The Economic Impact of Canola on the Canadian Econom: 2020 Update. Available online: https://www.canolacouncil.org/download/131/economic-impact/17818/economic-impact-report-canada_december-2020 (accessed on 10 October 2021).

57. Goering, K.J.; Eslick, R.; Brelsford, D.L. A search for high erucic acid containing oils in the cruciferae. *Econ. Bot.* **1965**, *19*, 251–256. [CrossRef]

58. Calhoun, W.; Crane, J.M.; Stamp, D.L. Development of a low glucosinolate, high erucic acid rapeseed breeding program. *J. Am. Oil Chem. Soc.* **1975**, *52*, 363–365. [CrossRef]

59. Tang, M.; Zhang, Y.; Liu, Y.; Tong, C.; Cheng, X.; Zhu, W.; Li, Z.; Huang, J.; Liu, S. Mapping loci controlling fatty acid profiles, oil and protein content by genome-wide association study in *Brassica napus*. *Crop J.* **2019**, *7*, 217–226. [CrossRef]

60. Chung, J.; Babka, H.L.; Graef, G.L.; Staswick, P.E.; Lee, D.J.; Cregan, P.B.; Shoemaker, R.C.; Specht, J.E. The seed protein, oil, and yield QTL on soybean linkage group I. (Cell Biology & Molecular Genetics).PDF. *Crop Sci.* **2003**, *43*, 1053–1067. [CrossRef]

61. Hu, Z.Y.; Hua, W.; Zhang, L.; Deng, L.B.; Wang, X.F.; Liu, G.H.; Hao, W.J.; Wang, H.Z. Seed structure characteristics to form ultrahigh oil content in rapeseed. *PLoS ONE* **2013**, *8*, 2–11. [CrossRef]

62. Ohlson, R. Modern processing of rapeseed. *J. Am. Oil Chem. Soc.* **1992**, *69*, 195–198. [CrossRef]

63. McCurdy, S.M. Effects of processing on the functional properties of canola/rapeseed protein. *J. Am. Oil Chem. Soc.* **1990**, *67*, 281–284. [CrossRef]

64. Mupondwa, E.; Li, X.; Wanasundara, J.P.D. Technoeconomic Prospects for Commercialization of Brassica (Cruciferous) Plant Proteins. *JAOCS, J. Am. Oil Chem. Soc.* **2018**, *95*, 903–922. [CrossRef]

65. Unger, E.H. Crushing and Extraction. *Canola Chem. Prod. Process. Util.* **2011**, 163–188.

66. Mag, T.K. Canola oil processing in Canada. *J. Am. Oil Chem. Soc.* **1983**, *60*, 380–384. [CrossRef]

67. Bailey, H.M.; Stein, H.H. Can the digestible indispensable amino acid score methodology decrease protein malnutrition. *Anim. Front.* **2019**, *9*, 18–23. [CrossRef]

68. Le Thanh, B.V.; Beltranena, E.; Zhou, X.; Wang, L.F.; Zijlstra, R.T. Amino acid and energy digestibility of *Brassica napus* canola meal from different crushing plants fed to ileal-cannulated grower pigs. *Anim. Feed Sci. Technol.* **2019**, *252*, 83–91. [CrossRef]

69. Foegeding, E.A.; Davis, J.P. Food protein functionality: A comprehensive approach. *Food Hydrocoll.* **2011**, *25*, 1853–1864. [CrossRef]

70. Tandang-Silvas, M.R.G.; Tecson-Mendoza, E.M.; Mikami, B.; Utsumi, S.; Maruyama, N. Molecular design of seed storage proteins for enhanced food physiochemical properties. *Annu. Rev. Food Sci. Technol.* **2011**, *2*, 59–73. [CrossRef] [PubMed]

71. Perera, S.P.; McIntosh, T.C.; Wanasundara, J.P.D. Structural properties of cruciferin and napin of *Brassica napus* (canola) show distinct responses to changes in pH and temperature. *Plants* **2016**, *5*, 36. [CrossRef]

72. Salazar-Villanea, S.; Bruininx, E.M.A.M.; Gruppen, H.; Carré, P.; Quinsac, A.; van der Poel, A.F.B. Effects of toasting time on digestive hydrolysis of soluble and insoluble 00-rapeseed meal proteins. *JAOCS J. Am. Oil Chem. Soc.* **2017**, *94*, 619–630. [CrossRef]

73. Wanasundara, J.P.D.; Abeysekara, S.J.; McIntosh, T.C.; Falk, K.C. Solubility differences of major storage proteins of brassicaceae oilseeds. *JAOCS, J. Am. Oil Chem. Soc.* **2012**, *89*, 869–881. [CrossRef]

74. Fukushima, D. Denaturation of Soybean proteins by organic solvents. *Cereal Chem.* **1969**, *46*, 156–163.

75. Thacker, P.A.; Newkirk, R.W. Performance of growing-finishing pigs fed barley-based diets containing toasted or non-toasted canola meal. *Can. J. Anim. Sci.* **2011**, *85*, 53–59. [CrossRef]

76. Mosenthin, R.; Messerschmidt, U.; Sauer, N.; Carré, P.; Quinsac, A.; Schöne, F. Effect of the desolventizing/toasting process on chemical composition and protein quality of rapeseed meal. *J. Anim. Sci. Biotechnol.* **2016**, *7*, 1–12. [CrossRef]

77. Newkirk, R.W.; Classen, H.L. The effects of toasting canola meal on body weight, feed conversion efficiency, and mortality in broiler chickens. *Poult. Sci.* **2002**, *81*, 815–825. [CrossRef]

78. Khattab, R.Y.; Arntfield, S.D. Functional properties of raw and processed canola meal. *LWT Food Sci. Technol.* **2009**, *42*, 1119–1124. [CrossRef]

79. He, R.; He, H.Y.; Chao, D.; Ju, X.; Aluko, R. Effects of High Pressure and Heat Treatments on Physicochemical and Gelation Properties of Rapeseed Protein Isolate. *Food Bioprocess Technol.* **2014**, *7*, 1344–1353. [CrossRef]

80. Lund, M.N.; Ray, C.A. Control of Maillard reactions in foods: Strategies and chemical mechanisms. *J. Agric. Food Chem.* **2017**, *65*, 4537–4552. [CrossRef] [PubMed]

81. Quality of Western Canadian Canola 2018. Available online: https://www.grainscanada.gc.ca/en/grain-research/export-quality/oilseeds/canola/2018/pdf/canola-quality-report-18.pdf (accessed on 10 October 2021).

82. Quality of Canadian Oilseed-Type Soybeans 2018. Available online: https://www.grainscanada.gc.ca/en/grain-research/export-quality/oilseeds/soybean-oil/2018/index.html (accessed on 10 October 2021).

83. Adewole, D.I.; Rogiewicz, A.; Dyck, B.; Slominski, B.A. Effects of canola meal source on the standardized ileal digestible amino acids and apparent metabolizable energy contents for broiler chickens. *Poult. Sci.* **2017**, *96*, 4298–4306. [CrossRef]

84. Kasprzak, M.M.; Houdijk, J.G.M.; Olukosi, O.A.; Appleyard, H.; Kightley, S.P.J.; Carré, P.; Sutton, T.; Wiseman, J. The content and standardized ileal digestibility of crude protein and amino acids in rapeseed co-products fed to pigs. *Livest. Sci.* **2018**, *208*, 22–27. [CrossRef]

85. Millward, D.J. Amino acid scoring patterns for protein quality assessment. *Br. J. Nutr.* **2012**, *108*. [CrossRef] [PubMed]

86. Schaafsma, G. The protein digestibility–corrected amino acid score. *J. Nutr.* **2000**, *130*, 1865S–1867S. [CrossRef]

87. Hughes, G.J.; Ryan, D.J.; Mukherjea, R.; Schasteen, C.S. Protein digestibility-corrected amino acid scores (PDCAAS) for soy protein isolates and concentrate: Criteria for evaluation. *J. Agric. Food Chem.* **2011**, *59*, 12707–12712. [CrossRef]

88. Bos, C.; Airinei, G.; Mariotti, F.; Benamouzig, R.; Bérot, S.; Evrard, J.; Fénart, E.; Tomé, D.; Gaudichon, C. The poor digestibility of rapeseed protein is balanced by its very high metabolic utilization in humans. *J. Nutr.* **2007**, *137*, 594–600. [CrossRef]

89. Fleddermann, M.; Fechner, A.; Rößler, A.; Bähr, M.; Pastor, A.; Liebert, F.; Jahreis, G. Nutritional evaluation of rapeseed protein compared to soy protein for quality, plasma amino acids, and nitrogen balance—A randomized cross-over intervention study in humans. *Clin. Nutr.* **2013**, *32*, 519–526. [CrossRef] [PubMed]

90. Gilani, G.S.; Cockell, K.A.; Sepehr, E. Effects of antinutritional factors on protein digestibility and amino acid availability in foods. *J. AOAC Int.* **2005**, *88*, 967–987. [CrossRef]

91. Bell, J.M. Factors affecting the nutritional value of canola meal: A review. *Can. J. Anim. Sci.* **1993**, *73*, 689–697. [CrossRef]

92. Hannoufa, A.; Pillai, B.V.S.; Chellamma, S. Genetic enhancement of *Brassica napus* seed quality. *Transgenic Res.* **2014**, *23*, 39–52. [CrossRef] [PubMed]

93. Harloff, H.J.; Lemcke, S.; Mittasch, J.; Frolov, A.; Wu, J.G.; Dreyer, F.; Leckband, G.; Jung, C. A mutation screening platform for rapeseed (*Brassica napus* L.) and the detection of sinapine biosynthesis mutants. *Theor. Appl. Genet.* **2012**, *124*, 957–969. [CrossRef]

94. Zhai, Y.; Yu, K.; Cai, S.; Hu, L.; Amoo, O.; Xu, L.; Yang, Y.; Ma, B.; Jiao, Y.; Zhang, C.; et al. Targeted mutagenesis of BnTT8 homologs controls yellow seed coat development for effective oil production in *Brassica napus* L. *Plant Biotechnol. J.* **2019**, *18*, pbi.13281. [CrossRef]

95. Sashidhar, N.; Harloff, H.; Jung, C. Identification of phytic acid mutants in oilseed rape (*Brassica napus*) by large scale screening of mutant populations through amplicon sequencing. *New Phytol.* **2019**, *225*, 2022–2034. [CrossRef] [PubMed]

96. Bérot, S.; Compoint, J.P.; Larré, C.; Malabat, C.; Guéguen, J. Large scale purification of rapeseed proteins (*Brassica napus* L.). *J. Chromatogr. B Anal. Technol. Biomed. Life Sci.* **2005**, *818*, 35–42. [CrossRef] [PubMed]

97. Steeves, T.A. The evolution and biological significance of seeds. *Can. J. Bot.* **1983**, *61*, 3550–3560. [CrossRef]

98. Crouch, M.L.; Sussex, I.M. Development and storage-protein synthesis in Brassica napus L. embryos in vivo and in vitro. *Planta* **1981**, *153*, 64–74. [CrossRef] [PubMed]

99. Norton, G.; Harris, J.F. Compositional changes in developing rape seed (*Brassica napus* L.). *Planta* **1975**, *123*, 163–174. [CrossRef] [PubMed]

100. Lord, E.M.; Russell, S.D. The mechanisms of pollination and fertilization in plants. *Annu. Rev. Cell Dev. Biol.* **2002**, *18*, 81–105. [CrossRef] [PubMed]

101. Lau, S.; Slane, D.; Herud, O.; Kong, J.; Jürgens, G. Early embryogenesis in flowering plants: Setting up the basic body pattern. *Annu. Rev. Plant Biol.* **2012**, *63*, 483–506. [CrossRef]

102. Capron, A.; Chatfield, S.; Provart, N.; Berleth, T. Embryogenesis: Pattern formation from a single cell. *Arab. B.* **2009**, *7*, e0126. [CrossRef]

103. Tykarska, T. Rape embryogenesis I. The proembryo development. *Acta Soc. Bot. Pol.* **1976**, *45*, 3–16. [CrossRef]

104. Tykarska, T. Rape embryogenesis. II. Development of embryo proper. *Acta Soc. Bot. Pol.* **1979**, *48*, 391–421. [CrossRef]

105. Ilić-Grubor, K.; Attree, S.M.; Fowke, L.C. Comparative morphological study of zygotic and microspore-derived embryos of *Brassica napus* L. as revealed by scanning electron microscopy. *Ann. Bot.* **1998**, *82*, 157–165. [CrossRef]

106. Fernandez, D.E.; Turner, F.R.; Crouch, M.L. In situ localization of storage protein mRNAs in developing meristems of *Brassica napus* embryos. *Development* **1991**, *111*, 299–313. [CrossRef]

107. Tykarska, T. Rape embryogenesis. III. Embryo development in time. *Acta Soc. Bot. Pol.* **1980**, *49*, 369–385. [CrossRef]

108. Yeung, E.C.; Rahman, M.H.; Thorpe, T.A. Comparative development of zygotic and microspore-derived embryos in *Brassica napus* L. cv Topas. I. Histodifferentiation. *Int. J. Plant Sci.* **1996**, *157*, 27–39. [CrossRef]

109. Shewry, P.R.; Halford, N.G. Cereal seed storage proteins: Structures, properties and role in grain utilization. *J. Exp. Bot.* **2002**, *53*, 947–958. [CrossRef]

110. Wanasundara, J.P.D. Proteins of brassicaceae oilseeds and their potential as a plant protein source. *Crit. Rev. Food Sci. Nutr.* **2011**, *51*, 635–677. [CrossRef]

111. Galland, M.; Huguet, R.; Arc, E.; Cueff, G.; Job, D.; Rajjou, L. Dynamic proteomics emphasizes the importance of selective mRNA translation and protein turnover during Arabidopsis seed germination. *Mol. Cell. Proteom.* **2014**, *13*, 252–268. [CrossRef] [PubMed]

112. Herman, E.M.; Larkins, B.A. Protein storage bodies and vacuoles. *Plant Cell* **1999**, *11*, 601–613. [CrossRef] [PubMed]

113. Job, C.; Rajjou, L.; Lovigny, Y.; Belghazi, M.; Job, D.; Loïc, R.; Yoann, L.; Maya, B.; Dominique, J.; Lovigny, Y.; et al. Patterns of protein oxidation in Arabidopsis seeds and during germination. *Plant Physiol.* **2005**, *138*, 790–802. [CrossRef] [PubMed]

114. Bieker, S.; Riester, L.; Doll, J.; Franzaring, J.; Fangmeier, A.; Zentgraf, U. Nitrogen supply drives senescence-related seed storage protein expression in rapeseed leaves. *Genes* **2019**, *10*. [CrossRef] [PubMed]

115. Osborne, T.B.; Osbornb, T.B. *The Vegetable Proteins*, 2nd ed.; Plimmer, R.H.A., Hopkins, F.G., Eds.; Longmans, Green and Co.: London, UK, 1924.

116. Hundertmark, M.; Hincha, D.K. LEA (Late Embryogenesis Abundant) proteins and their encoding genes in *Arabidopsis thaliana*. *BMC Genomics* **2008**, *9*, 1–22. [CrossRef] [PubMed]

117. Jolivet, P.; Boulard, C.; Bellamy, A.; Larré, C.; Barre, M.; Rogniaux, H.; D'Andréa, S.; Chardot, T.; Nesi, N. Protein composition of oil bodies from mature *Brassica napus* seeds. *Proteomics* **2009**, *9*, 3268–3284. [CrossRef]

118. Kubala, S.; Garnczarska, M.; Wojtyla, Ł.; Clippe, A.; Kosmala, A.; Żmieńko, A.; Lutts, S.; Quinet, M. Deciphering priming-induced improvement of rapeseed (*Brassica napus* L.) germination through an integrated transcriptomic and proteomic approach. *Plant Sci.* **2015**, *231*, 94–113. [CrossRef] [PubMed]

119. Yin, X.; He, D.; Gupta, R.; Yang, P. Physiological and proteomic analyses on artificially aged *Brassica napus* seed. *Front. Plant Sci.* **2015**, *6*, 1–11. [CrossRef] [PubMed]

120. Hajduch, M.; Casteel, J.E.; Hurrelmeyer, K.E.; Song, Z.; Agrawal, G.K.; Thelen, J.J. Proteomic analysis of seed filling in *Brassica napus*. Developmental characterization of metabolic isozymes using high-resolution two-dimensional gel electrophoresis. *Plant Physiol.* **2006**, *141*, 32–46. [CrossRef]

121. Borisjuk, L.; Neuberger, T.; Schwender, J.; Heinzel, N.; Sunderhaus, S.; Fuchs, J.; Hay, J.O.; Tschiersch, H.; Braun, H.-P.; Denolf, P.; et al. Seed architecture shapes embryo metabolism in oilseed rape. *Plant Cell* **2013**, *25*, 1625–1640. [CrossRef] [PubMed]

122. Hoglund, A.-S.; Rodin, J.; Larsson, E.; Rask, L.; Höglund, A.S.; Rödin, J.; Larsson, E.; Rask, L. Distribution of napin and cruciferin in developing rape seed embryos. *Plant Physiol.* **1992**, *98*, 509–515. [CrossRef]

123. Sjödahl, S.; Gustavsson, H.O.; Rödin, J.; Lenman, M.; Höglund, A.S.; Rask, L. Cruciferin gene families are expressed coordinately but with tissue-specific differences during *Brassica napus* seed development. *Plant Mol. Biol.* **1993**, *23*, 1165–1176. [CrossRef]

124. Tandang-Silvas, M.R.G.; Fukuda, T.; Fukuda, C.; Prak, K.; Cabanos, C.; Kimura, A.; Itoh, T.; Mikami, B.; Utsumi, S.; Maruyama, N. Conservation and divergence on plant seed 11S globulins based on crystal structures. *Biochim. Biophys. Acta Proteins Proteomics* **2010**, *1804*, 1432–1442. [CrossRef]

125. Nietzel, T.; Dudkina, N.V.; Haase, C.; Denolf, P.; Semchonok, D.A.; Boekema, E.J.; Braun, H.P.; Sunderhaus, S. The native structure and composition of the cruciferin complex in *Brassica napus*. *J. Biol. Chem.* **2013**, *288*, 2238–2245. [CrossRef]

126. Müntz, K. Deposition of storage proteins. *Plant Mol. Biol.* **1998**, *38*, 77–99. [CrossRef] [PubMed]

127. Jung, R.; Paul Scott, M.; Nam, Y.W.; Beaman, T.W.; Bassüner, R.; Saalbach, I.; Müntz, K.; Nielsen, N.C. The role of proteolysis in the processing and assembly of 11S seed globulins. *Plant Cell* **1998**, *10*, 343–357. [CrossRef] [PubMed]

128. Aider, M.; Barbana, C. Canola proteins: Composition, extraction, functional properties, bioactivity, applications as a food ingredient and allergenicity—A practical and critical review. *Trends Food Sci. Technol.* **2011**, *22*, 21–39. [CrossRef]

129. Kim, J.H.; Varankovich, N.V.; Stone, A.K.; Nickerson, M.T. Nature of protein-protein interactions during the gelation of canola protein isolate networks. *Food Res. Int.* **2016**, *89*, 408–414. [CrossRef]

130. Pudel, F.; Tressel, R.P.; Düring, K. Production and properties of rapeseed albumin. *Lipid Technol.* **2015**, *27*, 112–114. [CrossRef]

131. Byczyńska, A.; Barciszewski, J. The biosynthesis, structure and properties of napin—The storage protein from rape seeds. *J. Plant Physiol.* **1999**, *154*, 417–425. [CrossRef]

132. Lönnerdal, B.; Janson, J.-C. Studies on Brassica seed proteins. *Biochim. Biophys. Acta Protein Struct.* **1972**, *278*, 175–183. [CrossRef]

133. Ericson, M.L.; Rodin, J.; Lenman, M.; Glimelius, K.; Josefsson, L.G.; Rask, L.; Rödin, J.; Lenman, M.; Glimelius, K.; Josefsson, L.G.; et al. Structure of the rapeseed 1.7 S storage protein, napin, and its precursor. *J. Biol. Chem.* **1986**, *261*, 14576–14581. [CrossRef]

134. Mylne, J.S.; Hara-Nishimura, I.; Rosengren, K.J. Seed storage albumins: Biosynthesis, trafficking and structures. *Funct. Plant Biol.* **2014**, *41*, 671–677. [CrossRef]

135. Boutilier, K.; Hattori, J.; Baum, B.R.; Miki, B.L. Evolution of 2S albumin seed storage protein genes in the Brassicaceae. *Biochem. Syst. Ecol.* **1999**, *27*, 223–234. [CrossRef]

136. Josefsson, L.G.; Lenman, M.; Ericson, M.L.; Rask, L. Structure of a gene encoding the 1.7 S storage protein, napin, from *Brassica napus*. *J. Biol. Chem.* **1987**, *262*, 12196–12201. [CrossRef]

137. Scofield, S.R.; Crouch, M.L. Nucleotide sequence of a member of the napin storage protein family from *Brassica napus*. *J. Biol. Chem.* **1987**, *262*, 12202–12208. [CrossRef]

138. José-Estanyol, M.; Gomis-Rüth, F.X.; Puigdomènech, P. The eight-cysteine motif, a versatile structure in plant proteins. *Plant Physiol. Biochem.* **2004**, *42*, 355–365. [CrossRef]

139. Malabat, C.; nchez-Vioque, R.I.S.; Rabiller, C.; Gu guen, J. Emulsifying and foaming properties of native and chemically modified peptides from the 2S and 12S proteins of rapeseed (Brassica napus L.). *J. Am. Oil Chem. Soc.* **2001**, *78*, 235–242. [CrossRef]

140. Schwartz, J.M.; Solé, V.; Guéguen, J.; Ropers, M.H.; Riaublanc, A.; Anton, M. Partial replacement of β-casein by napin, a rapeseed protein, as ingredient for processed foods: Thermoreversible aggregation. *LWT Food Sci. Technol.* **2015**, *63*, 562–568. [CrossRef]

141. Malabat, C.; Atterby, H.; Chaudhry, Q.; Renard, M.; Guéguen, J.; De Recherche, U.; Végétales, P.; Kingdom, U.; Rheu, L. Genetic variability of rapeseed protein composition. In Proceedings of 11th International Rapeseed Congress, Royal Veterinary and Agricultural University, Frederiksberg, Denmark, 6 July 2003.

142. Kawakatsu, T.; Takaiwa, F. Cereal seed storage protein synthesis: Fundamental processes for recombinant protein production in cereal grains. *Plant Biotechnol. J.* **2010**, *8*, 939–953. [CrossRef]

143. Le Signor, C.; Aimé, D.; Bordat, A.; Belghazi, M.; Labas, V.; Gouzy, J.; Young, N.D.; Prosperi, J.M.; Leprince, O.; Thompson, R.D.; et al. Genome-wide association studies with proteomics data reveal genes important for synthesis, transport and packaging of globulins in legume seeds. *New Phytol.* **2017**, *214*, 1597–1613. [CrossRef] [PubMed]

144. Members of the Complex Trait Consortium The nature and identification of quantitative trait loci: A community's view. *Nat. Rev. Genet.* **2003**, *4*, 911–916. [CrossRef]

145. Zhao, J.; Becker, H.C.; Zhang, D.; Zhang, Y.; Ecke, W. Conditional QTL mapping of oil content in rapeseed with respect to protein content and traits related to plant development and grain yield. *Theor. Appl. Genet.* **2006**, *113*, 33–38. [CrossRef]

146. Würschum, T.; Liu, W.; Maurer, H.P.; Abel, S.; Reif, J.C. Dissecting the genetic architecture of agronomic traits in multiple segregating populations in rapeseed (Brassica napus L.). *Theor. Appl. Genet.* **2012**, *124*, 153–161. [CrossRef]

147. Teh, L.; Möllers, C. Genetic variation and inheritance of phytosterol and oil content in a doubled haploid population derived from the winter oilseed rape Sansibar × Oase cross. *Theor. Appl. Genet.* **2016**, *129*, 181–199. [CrossRef] [PubMed]

148. Behnke, N.; Suprianto, E.; Möllers, C. A major QTL on chromosome C05 significantly reduces acid detergent lignin (ADL) content and increases seed oil and protein content in oilseed rape (*Brassica napus* L.). *Theor. Appl. Genet.* **2018**, *131*, 2477–2492. [CrossRef] [PubMed]

149. Chao, H.; Wang, H.; Wang, X.; Guo, L.; Gu, J.; Zhao, W.; Li, B.; Chen, D.; Raboanatahiry, N.; Li, M. Genetic dissection of seed oil and protein content and identification of networks associated with oil content in *Brassica napus*. *Sci. Rep.* **2017**, *7*, 1–16. [CrossRef]

150. Chen, F.; Zhang, W.; Yu, K.; Sun, L.; Gao, J.; Zhou, X.; Peng, Q.; Fu, S.; Hu, M.; Long, W.; et al. Unconditional and conditional QTL analyses of seed fatty acid composition in *Brassica napus* L. *BMC Plant Biol.* **2018**, *18*, 1–14. [CrossRef] [PubMed]

151. Bao, B.; Chao, H.; Wang, H.; Zhao, W.; Zhang, L.; Raboanatahiry, N.; Wang, X.; Wang, B.; Jia, H.; Li, M. Stable, environmental specific and novel QTL identification as well as genetic dissection of fatty acid metabolism in *Brassica napus*. *Front. Plant Sci.* **2018**, *9*, 1–25. [CrossRef]

152. Toroser, D.; Thormann, C.E.; Osborn, T.C.; Mithen, R. RFLP mapping of quantitative trait loci controlling seed aliphatic-glucosinolate content in oilseed rape (*Brassica napus* L.). *Theor. Appl. Genet.* **1995**, *91*, 802–808. [CrossRef]

153. Uzunova, M.; Ecke, W.; Weissleder, K.; Röbbelen, G. Mapping the genome of rapeseed (*Brassica napus* L.). I. Construction of an RFLP linkage map and localization of QTLs for seed glucosinolate content. *Theor. Appl. Genet.* **1995**, *90*, 194–204. [CrossRef]

154. Howell, P.M.; Sharpe, A.G.; Lydiate, D.J. Homoeologous loci control the accumulation of seed glucosinolates in oilseed rape (*Brassica napus*). *Genome* **2003**, *46*, 454–460. [CrossRef]

155. Schatzki, J.; Ecke, W.; Becker, H.C.; Möllers, C. Mapping of QTL for the seed storage proteins cruciferin and napin in a winter oilseed rape doubled haploid population and their inheritance in relation to other seed traits. *Theor. Appl. Genet.* **2014**, *127*, 1213–1222. [CrossRef]

156. Wen, J.; Xu, J.F.; Long, Y.; Wu, J.G.; Xu, H.M.; Meng, J.L.; Shi, C.H. QTL mapping based on the embryo and maternal genetic systems for non-essential amino acids in rapeseed (*Brassica napus* L.) meal. *J. Sci. Food Agric.* **2016**, *96*, 465–473. [CrossRef]

157. Fauteux, F.; Strömvik, M.V. Seed storage protein gene promoters contain conserved DNA motifs in Brassicaceae, Fabaceae and Poaceae. *BMC Plant Biol.* **2009**, *9*, 1–11. [CrossRef]

158. Sjödahl, S.; Gustavsson, H.O.; Rödin, J.; Rask, L. Deletion analysis of the Brassica napus cruciferin gene *cru 1* promoter in transformed tobacco: Promoter activity during early and late stages of embryogenesis is influenced by *cis*-acting elements in partially separate regions. *Planta* **1995**, *197*, 264–271. [CrossRef] [PubMed]

159. Stålberg, K.; Ellerstöm, M.; Ezcurra, I.; Ablov, S.; Rask, L. Disruption of an overlapping E-box/ABRE motif abolished high transcription of the *napA* storage-protein promoter in transgenic *Brassica napus* seeds. *Planta* **1996**, *199*, 515–519. [CrossRef]

160. Parcy, F.; Valon, C.; Raynal, M.; Gaubier-Comella, P.; Delseny, M.; Giraudat, J. Regulation of gene expression programs during Arabidopsis seed development: Roles of the *ABI3* locus and of endogenous abscisic acid. *Plant Cell* **1994**, *6*, 1567–1582. [CrossRef]

161. Kroj, T.; Savino, G.; Valon, C.; Giraudat, J.; Parcy, F. Regulation of storage protein gene expression in Arabidopsis. *Development* **2003**, *130*, 6065–6073. [CrossRef]

162. Lara, P.; Oñate-Sánchez, L.; Abraham, Z.; Ferrándiz, C.; Díaz, I.; Carbonero, P.; Vicente-Carbajosa, J. Synergistic activation of seed storage protein gene expression in *Arabidopsis* by ABI3 and two bZIPs related to OPAQUE2. *J. Biol. Chem.* **2003**, *278*, 21003–21011. [CrossRef] [PubMed]

163. Ezcurra, I.; Wycliffe, P.; Nehlin, L.; Ellerström, M.; Rask, L. Transactivation of the *Brassica napus* napin promoter by ABI3 requires interaction of the conserved B2 and B3 domains of ABI3 with different cis-elements: B2 mediates activation through an ABRE, whereas B3 interacts with an RY/G-box. *Plant J.* **2000**, *24*, 57–66. [CrossRef] [PubMed]

164. Vicient, C.M.; Bies-Etheve, N.; Delseny, M. Changes in gene expression in the *leafy cotyledon1* (*lec1*) and *fusca3* (*fus3*) mutants of *Arabidopsis thaliana* L. *J. Exp. Bot.* **2000**, *51*, 995–1003. [CrossRef]

165. Kagaya, Y.; Toyoshima, R.; Okuda, R.; Usui, H.; Yamamoto, A.; Hattori, T. LEAFY COTYLEDON1 controls seed storage protein genes through its regulation of *FUSCA3* and *ABSCISIC ACID INSENSITIVE3*. *Plant Cell Physiol.* **2005**, *46*, 399–406. [CrossRef]

166. Pelletier, J.M.; Kwong, R.W.; Park, S.; Le, B.H.; Baden, R.; Cagliari, A.; Hashimoto, M.; Munoz, M.D.; Fischer, R.L.; Goldberg, R.B.; et al. LEC1 sequentially regulates the transcription of genes involved in diverse developmental processes during seed development. *Proc. Natl. Acad. Sci. USA* **2017**, *114*, E6710–E6719. [CrossRef]

167. Gao, C.; Qi, S.; Liu, K.; Li, D.; Jin, C.; Li, Z.; Huang, G.; Hai, J.; Zhang, M.; Chen, M. MYC2, MYC3, and MYC4 function redundantly in seed storage protein accumulation in *Arabidopsis*. *Plant Physiol. Biochem.* **2016**, *108*, 63–70. [CrossRef]

168. Verdier, J.; Thompson, R.D. Transcriptional regulation of storage protein synthesis during dicotyledon seed filling. *Plant Cell Physiol.* **2008**, *49*, 1263–1271. [CrossRef] [PubMed]

169. Sah, S.K.; Reddy, K.R.; Li, J. Abscisic acid and abiotic stress tolerance in crop plants. *Front. Plant Sci.* **2016**, *7*, 1–26. [CrossRef] [PubMed]

170. Barthet, V.J. Comparison between Canadian canola harvest and export surveys. *Plants* **2016**, *5*, 1209–1214. [CrossRef]

171. Triboï, E.; Martre, P.; Triboï-Blondel, A.M.; Croue, S. De Environmentally-induced changes in protein composition in developing grains of wheat are related to changes in total protein content. *J. Exp. Bot.* **2003**, *54*, 1731–1742. [CrossRef]

172. Staswick, P.E.; Huang, J.F.; Rhee, Y. Nitrogen and methyl jasmonate induction of soybean vegetative storage protein genes. *Plant Physiol.* **1991**, *96*, 130–136. [CrossRef] [PubMed]

173. Paek, N.C.; Sexton, P.J.; Naeve, S.L.; Shibles, R. Differential accumulation of soybean seed storage protein subunits in response to sulfur and nitrogen nutritional sources. *Plant Prod. Sci.* **2000**, *3*, 268–274. [CrossRef]

174. Borpatragohain, P.; Rose, T.J.; King, G.J. Fire and brimstone: Molecular interactions between sulfur and glucosinolate biosynthesis in model and crop brassicaceae. *Front. Plant Sci.* **2016**, *7*, 1–18. [CrossRef]

175. Wang, Z.-H.; Li, S.-X.; Malhi, S. Effects of fertilization and other agronomic measures on nutritional quality of crops. *J. Sci. Food Agric.* **2008**, *88*, 7–23. [CrossRef]

176. Grant, C.A.; Clayton, G.W.; Johnston, A.M. Sulphur fertilizer and tillage effects on canola seed quality in the Black soil zone of western Canada. *Can. J. Plant Sci.* **2003**, *83*, 745–758. [CrossRef]

177. Poisson, E.; Trouverie, J.; Brunel-Muguet, S.; Akmouche, Y.; Pontet, C.; Pinochet, X.; Avice, J.C. Seed yield components and seed quality of oilseed rape are impacted by sulfur fertilization and its interactions with nitrogen fertilization. *Front. Plant Sci.* **2019**, *10*, 1–14. [CrossRef]

178. Chardigny, J.-M.M.; Walrand, S. Plant protein for food: Opportunities and bottlenecks. *OCL Oilseeds fats, Crop. Lipids* **2016**, *23*, D404. [CrossRef]

179. Mejia, L.A.; Korgaonkar, C.K.; Schweizer, M.; Chengelis, C.; Marit, G.; Ziemer, E.; Grabiel, R.; Empie, M. A 13-week sub-chronic dietary toxicity study of a cruciferin-rich canola protein isolate in rats. *Food Chem. Toxicol.* **2009**, *47*, 2645–2654. [CrossRef] [PubMed]

180. Mejia, L.A.; Korgaonkar, C.K.; Schweizer, M.; Chengelis, C.; Novilla, M.; Ziemer, E.; Williamson-Hughes, P.S.; Grabiel, R.; Empie, M. A 13-week dietary toxicity study in rats of a napin-rich canola protein isolate. *Regul. Toxicol. Pharmacol.* **2009**, *55*, 394–402. [CrossRef] [PubMed]

181. Masilamani, M.; Commins, S.; Shreffler, W. Determinants of food allergy. *Immunol. Allergy Clin. North Am.* **2012**, *32*, 11–33. [CrossRef]

182. Moreno, F.J.; Clemente, A. 2S Albumin Storage Proteins: What Makes them Food Allergens? *Open Biochem. J.* **2008**, *2*, 16–28. [CrossRef]

183. Sicherer, S.H.; Sampson, H.A. Food allergy: A review and update on epidemiology, pathogenesis, diagnosis, prevention, and management. *J. Allergy Clin. Immunol.* **2018**, *141*, 41–58. [CrossRef]

184. Waserman, S.; Bégin, P.; Watson, W. IgE-mediated food allergy. *Clin. Rev. Allergy Immunol.* **2019**, *57*, 244–260. [CrossRef]

185. Gould, H.J.; Sutton, B.J. IgE in allergy and asthma today. *Nat. Rev. Immunol.* **2008**, *8*, 205–217. [CrossRef]

186. Monsalve, R.I.; Gonzalez de la Pena, M.A.; Lopez-Otiin, C.; Fiandor, A.; Fernandez, C.; Villalba, M.; Rodriguez, R. Detection, isolation and complete amino acid sequence of an aeroallergenic protein from rapeseed flour. *Clin Exp Allergy* **1997**, *27*, 833–841. [CrossRef]

187. Puumalainen, T.J.; Puustinen, A.; Poikonen, S.; Turjanmaa, K.; Palosuo, T.; Vaali, K. Proteomic identification of allergenic seed proteins, napin and cruciferin, from cold-pressed rapeseed oils. *Food Chem.* **2015**, *175*, 381–385. [CrossRef] [PubMed]

188. EFSA Panel on Dietetic Products Nutrition and Allergies Scientific Opinion on the safety of "rapeseed protein isolate" as a Novel Food ingredient. *EFSA J.* **2013**, *11*, 1–23. [CrossRef]

189. Fiocchi, A.; Dahdah, L.; Riccardi, C.; Mazzina, O.; Fierro, V. Preacautionary labelling of cross-reactive foods: The case of rapeseed. *Asthma Res. Pract.* **2016**, *2*, 1–8. [CrossRef] [PubMed]

190. Bonds, R.S.; Midoro-Horiuti, T.; Goldblum, R. A structural basis for food allergy: The role of cross-reactivity. *Curr. Opin. Allergy Clin. Immunol.* **2008**, *8*, 82–86. [CrossRef] [PubMed]

191. L'Hocine, L.; Pitre, A.; Pitre, M. Detection and identification of allergens from Canadian mustard varieties of *Sinapis alba* and *Brassica juncea*. *Biomolecules* **2019**, *9*, 489. [CrossRef] [PubMed]

192. Government of Canada Regulations amending the food and drug regulations (1220—Enhanced labeling for food allergen and gluten sources and added sulphites). *Canada Gaz. Part II* **2011**, *145*, 325.

193. Mustard: A Priority Food Allergen in Canada 2009 . Available online: https://www.canada.ca/en/health-canada/services/food-nutrition/reports-publications/food-labelling/mustard-priority-food-allergen-canada-systematic-review.html (accessed on 10 October 2021).

194. Marambe, H.K.; McIntosh, T.C.; Cheng, B.; Wanasundara, J.P.D.D. Quantification of major 2S allergen protein of yellow mustard using anti-Sin a 1 epitope antibody. *Food Control* **2014**, *44*, 233–241. [CrossRef]
195. Gylling, H. Rapeseed oil does not cause allergic reactions. *Allergy Eur. J. Allergy Clin. Immunol.* **2006**, *61*, 895. [CrossRef]
196. Paul, R.H.; Frestedt, J.; Magurany, K. GRAS from the ground up: Review of the Interim Pilot Program for GRAS notification. *Food Chem. Toxicol.* **2017**, *105*, 140–150. [CrossRef] [PubMed]
197. van der Spiegel, M.; Noordam, M.Y.; van der Fels-Klerx, H.J. Safety of novel protein sources (insects, microalgae, seaweed, duckweed, and rapeseed) and legislative aspects for their application in food and feed production. *Compr. Rev. Food Sci. Food Saf.* **2013**, *12*, 662–678. [CrossRef]
198. Day, L. Proteins from land plants—Potential resources for human nutrition and food security. *Trends Food Sci. Technol.* **2013**, *32*, 25–42. [CrossRef]
199. Kim, S.W.; Less, J.F.; Wang, L.; Yan, T.; Kiron, V.; Kaushik, S.J.; Lei, X.G. Meetin global feed protein demand: Challenge, opportunity, and strategy. *Annu. Rev. Anim. Biosci.* **2019**, *7*, 221–243. [CrossRef]
200. Lomascolo, A.; Uzan-Boukhris, E.; Sigoillot, J.C.; Fine, F. Rapeseed and sunflower meal: A review on biotechnology status and challenges. *Appl. Microbiol. Biotechnol.* **2012**, *95*, 1105–1114. [CrossRef]
201. Thomke, S. Review of rapeseed meal in animal nutrition: Ruminant animals. *J. Am. Oil Chem. Soc.* **1981**, *58*, 805–810. [CrossRef]
202. Swanepoel, N.; Robinson, P.H.; Erasmus, L.J. Determining the optimal ratio of canola meal and high protein dried distillers grain protein in diets of high producing Holstein dairy cows. *Anim. Feed Sci. Technol.* **2014**, *189*, 41–53. [CrossRef]
203. Huhtanen, P.; Hetta, M.; Swensson, C. Evaluation of canola meal as a protein supplement for dairy cows: A review and a meta-analysis. *Can. J. Anim. Sci.* **2012**, *91*, 529–543. [CrossRef]
204. He, M.L.; Gibb, D.; McKinnon, J.J.; McAllister, T.A. Effect of high dietary levels of canola meal on growth performance, carcass quality and meat fatty acid profiles of feedlot cattle. *Can. J. Anim. Sci.* **2013**, *93*, 269–280. [CrossRef]
205. Mejicanos, G.; Sanjayan, N.; Kim, I.H.; Nyachoti, C.M. Recent advances in canola meal utilization in swine nutrition. *J. Anim. Sci. Technol.* **2016**, *58*, 1–13. [CrossRef] [PubMed]
206. Landero, J.L.; Beltranena, E.; Zijlstra, R.T. Growth performance and preference studies to evaluate solvent-extracted *Brassica napus* or *Brassica juncea* canola meal fed to weaned pigs. *J. Anim. Sci.* **2012**, *90*, 406–408. [CrossRef] [PubMed]
207. Chai, S.J.; Cole, D.; Nisler, A.; Mahon, B.E. Poultry: The most common food in outbreaks with known pathogens, United States, 1998-2012. *Epidemiol. Infect.* **2017**, *145*, 316–325. [CrossRef] [PubMed]
208. Hossain, M.A.A.; Islam, A.F.F.; Iji, P.A.A. Energy utilization and performance of broiler chickens raised on diets with vegetable proteins or conventional feeds. *Asian J. Poult. Sci.* **2012**, *6*, 117–128. [CrossRef]
209. Ravindran, V.; Abdollahi, M.R.; Bootwalla, S.M. Nutrient analysis, metabolizable energy, and digestible amino acids of soybean meals of different origins for broilers. *Poult. Sci.* **2014**, *93*, 2567–2577. [CrossRef] [PubMed]
210. Vieira, S.L.; Lima, I.L. Live performance, water intake and excreta characteristics of broilers fed all vegetable diets based on corn and soybean meal. *Int. J. Poult. Sci.* **2005**, *4*, 365–368. [CrossRef]
211. Kasprzak, M.M.; Houdijk, J.G.M.; Liddell, S.; Davis, K.; Olukosi, O.A.; Kightley, S.; White, G.A.; Wiseman, J. Rapeseed napin and cruciferin are readily digested by poultry. *J. Anim. Physiol. Anim. Nutr. (Berl).* **2017**, *101*, 658–666. [CrossRef]
212. Proudfoot, F.C.; Hulan, H.W.; McRae, K.B. Effect of feeding poultry diets supplemented with rapeseed meal as a primary protein source to juvenile and adult meat breeder genotypes. *Can. J. Anim. Sci.* **1983**, *63*, 957–965. [CrossRef]
213. Bryan, D.D.S.L.; Abbott, D.A.; Van Kessel, A.G.; Classen, H.L. In vivo digestion characteristics of protein sources fed to broilers. *Poult. Sci.* **2019**, *98*, 3313–3325. [CrossRef] [PubMed]
214. The State of World Fisheries and Aquaculture 2018 . Available online: https://www.fao.org/policy-support/tools-and-publications/resources-details/en/c/1269332/ (accessed on 10 October 2021).
215. Turchini, G.M.; Trushenski, J.T.; Glencross, B.D. Throughts for the future of aquaculture nutrition: REaligning perspectives to reflect contemporary issues related to judicious use of marine resources in aquafeeds. *N. Am. J. Aquac.* **2019**, *81*, 13–39. [CrossRef]
216. Adem, H.N.; Tressel, R.P.; Pudel, F.; Slawski, H.; Schulz, C. Rapeseed use in aquaculture. *OCL Oilseeds fats* **2014**, *21*, 1–9. [CrossRef]
217. Nagel, F.; Slawski, H.; Adem, H.; Tressel, R.P.; Wysujack, K.; Schulz, C. Albumin and globulin rapeseed protein fractions as fish meal alternative in diets fed to rainbow trout (Oncorhynchus mykiss W.). *Aquaculture* **2012**, *354–355*, 121–127. [CrossRef]
218. Slawski, H.; Adem, H.; Tressel, R.P.; Wysujack, K.; Koops, U.; Kotzamanis, Y.; Wuertz, S.; Schulz, C. Total fish meal replacement with rapeseed protein concentrate in diets fed to rainbow trout (*Oncorhynchus mykiss* Walbaum). *Aquac. Int.* **2012**, *20*, 443–453. [CrossRef]
219. Luo, Z.; Li, X.D.; Wang, W.M.; Tan, X.Y.; Liu, X. Partial replacement of fish meal by a mixture of soybean meal and rapeseed meal in practical diets for juvenile Chinese mitten crab *Eriocheir sinensis*: Effects on growth performance and in vivo digestibility. *Aquac. Res.* **2011**, *42*, 1615–1622. [CrossRef]
220. Suárez, J.A.; Gaxiola, G.; Mendoza, R.; Cadavid, S.; Garcia, G.; Alanis, G.; Suárez, A.; Faillace, J.; Cuzon, G. Substitution of fish meal with plant protein sources and energy budget for white shrimp *Litopenaeus vannamei* (Boone, 1931). *Aquaculture* **2009**, *289*, 118–123. [CrossRef]
221. Cruz-Suarez, L.E.; Ricque-Marie, D.; Tapia-Salazar, M.; McCallum, I.M.; Hickling, D. Assessment of differently processed feed pea (*Pisum sativum*) meals and canola meal (*Brassica sp.*) in diets for blue shrimp (*Litopenaeus stylirostris*). *Aquaculture* **2001**, *196*, 87–104. [CrossRef]

222. Kumar, P.; Chatli, M.K.; Mehta, N.; Singh, P.; Malav, O.P.; Verma, A.K. Meat analogues: Health promising sustainable meat substitutes. *Crit. Rev. Food Sci. Nutr.* **2017**, *57*, 923–932. [CrossRef] [PubMed]

223. Hoek, A.C.; Luning, P.A.; Weijzen, P.; Engels, W.; Kok, F.J.; de Graaf, C. Replacement of meat by meat substitutes. A survey on person- and product-related factors in consumer acceptance. *Appetite* **2011**, *56*, 662–673. [CrossRef] [PubMed]

224. Dekkers, B.L.; Boom, R.M.; van der Goot, A.J. Structuring processes for meat analogues. *Trends Food Sci. Technol.* **2018**, *81*, 25–36. [CrossRef]

225. Wu, J.; Muir, A.D. Comparative structural, emulsifying, and biological properties of 2 major canola proteins, cruciferin and napin. *J. Food Sci.* **2008**, *73*. [CrossRef]

226. Kohno-Murase, J.; Murase, M.; Ichikawa, H.; Imamura, J. Improvement in the quality of seed storage protein by transformation of *Brassica napus* with an antisense gene for cruciferin. *Theor. Appl. Genet.* **1995**, *91*, 627–631. [CrossRef]

227. Falco, S.C.; Guida, T.; Locke, M.; Mauvais, J.; Sanders, C.; Ward, R.T.; Webber, P. Transgenic Canola and Soyban Seeds with Increased Lysine. *Nat. Biotechnol.* **1995**, *13*, 577–582. [CrossRef]

228. Cheung, L.; Wanasundara, J.; Nickerson, M.T. The Effect of pH and NaCl Levels on the Physicochemical and Emulsifying Properties of a Cruciferin Protein Isolate. *Food Biophys.* **2014**, *9*, 105–113. [CrossRef]

229. Cheung, L.; Wanasundara, J.; Nickerson, M.T. Effect of pH and NaCl on the Emulsifying Properties of a Napin Protein Isolate. *Food Biophys.* **2014**, *10*, 30–38. [CrossRef]

230. Gerzhova, A.; Mondor, M.; Benali, M.; Aider, M. Study of total dry matter and protein extraction from canola meal as affected by the pH, salt addition and use of zeta-potential/turbidimetry analysis to optimize the extraction conditions. *Food Chem.* **2016**, *201*, 243–252. [CrossRef] [PubMed]

231. Congdon, R.W.; Muth, G.W.; Splittgerber, A.G. The binding interaction of Coomassie blue with proteins. *Anal. Biochem.* **1993**, *213*, 407–413. [CrossRef]

232. Towbin, H.; Staehelin, T.; Gordon, J. Electrophoretic transfer of proteins from polyacrylamide gels to nitrocellulose sheets: Procedure and some applications. *Proc. Natl. Acad. Sci. USA* **1979**, *76*, 4350–4354. [CrossRef] [PubMed]

233. Moritz, C.P. 40 years Western blotting: A scientific birthday toast. *J. Proteomics* **2020**, *212*. [CrossRef]

234. Janes, K.A. An analysis of critical factors for quantitative immunoblotting. *Sci. Signal.* **2015**, *8*, rs2. [CrossRef] [PubMed]

235. Rödin, J.; Rask, L. Characterization of the 12S storage protein of *Brassica napus* (cruciferin): Disulfide bonding between subunits. *Physiol. Plant.* **1990**, *79*, 421–426. [CrossRef]

236. Kawagoe, Y.; Suzuki, K.; Tasaki, M.; Yasuda, H.; Akagi, K.; Katoh, E.; Nishizawa, N.K.; Ogawa, M.; Takaiwa, F. The critical role of disulfide bond formation in protein sorting in the endosperm of rice. *Plant Cell* **2005**, *17*, 1141–1153. [CrossRef]

237. Wan, L.; Ross, A.R.S.; Yang, J.; Hegedus, D.D.; Kermode, A.R. Phosphorylation of the 12 S globulin cruciferin in wild-type and *abi1-1* mutant *Arabidopsis thaliana* (thale cress) seeds. *Biochem. J.* **2007**, *404*, 247–256. [CrossRef]

238. Murphy, D.J.; Cummins, I.; Kang, A.S. Synthesis of the major oil-body membrane protein in developing rapeseed (*Brassica napus*) embryos. Integration with storage-lipid and storage-protein synthesis and implications for the mechanism of oil-body formation. *Biochem. J.* **1989**, *258*, 285–293. [CrossRef]

239. Murphy, D.J.; Cummins, I. Biosynthesis of Seed Storage Products during Embryogenesis in Rapeseed, Brassica napus. *J. Plant Physiol.* **1989**, *135*, 63–69. [CrossRef]

240. Sikdar, M.S.I.; Bowra, S.; Schmidt, D.; Dionisio, G.; Holm, P.B.; Vincze, E. Targeted modification of storage protein content resulting in improved amino acid composition of barley grain. *Transgenic Res.* **2016**, *25*, 19–31. [CrossRef]

241. Gil-Humanes, J.; Pistón, F.; Giménez, M.J.; Martín, A.; Barro, F. The Introgression of RNAi Silencing of γ-Gliadins into Commercial Lines of Bread Wheat Changes the Mixing and Technological Properties of the Dough. *PLoS ONE* **2012**, *7*. [CrossRef]

242. Kim, H.J.; Lee, J.Y.; Yoon, U.H.; Lim, S.H.; Kim, Y.M. Effects of reduced prolamin on seed storage protein composition and the nutritional quality of rice. *Int. J. Mol. Sci.* **2013**, *14*, 17073–17084. [CrossRef]

243. Wu, Y.; Messing, J. Proteome balancing of the maize seed for higher nutritional value. *Front. Plant Sci.* **2014**, *5*, 1–5. [CrossRef]

244. Henderson, I.R.; Salt, D.E. Natural genetic variation and hybridization in plants. *J. Exp. Bot.* **2017**, *68*, 5415–5417. [CrossRef] [PubMed]

245. Carmona, S.; Alvarez, J.B.; Caballero, L. Genetic diversity for morphological traits and seed storage proteins in Spanish rivet wheat. *Biol. Plant.* **2010**, *54*, 69–75. [CrossRef]

246. Liu, S.; Ohta, K.; Dong, C.; Thanh, V.C.; Ishimoto, M.; Qin, Z.; Hirata, Y. Genetic diversity of soybean (*Glycine max* (L.) Merrill) 7S globulin protein subunits. *Genet. Resour. Crop Evol.* **2006**, *53*, 1209–1219. [CrossRef]

247. Kim, W.S.; Gillman, J.D.; Krishnan, H.B. Identification of a plant introduction soybean line with genetic lesions affecting two distinct glycinin subunits and evaluation of impacts on protein content and composition. *Mol. Breed.* **2013**, *32*, 291–298. [CrossRef]

248. Hu, Y.; Wu, G.; Cao, Y.; Wu, Y.; Xiao, L.; Li, X.; Lu, C. Breeding response of transcript profiling in developing seeds of *Brassica napus*. *BMC Mol. Biol.* **2009**, *10*. [CrossRef]

249. Rizzo, G.; Baroni, L. Soy, Soy Foods and Their Role in Vegetarian Diets. *Nutrients* **2018**, *10*, 43. [CrossRef]

250. Nishinari, K.; Fang, Y.; Guo, S.; Phillips, G.O. Soy proteins: A review on composition, aggregation and emulsification. *Food Hydrocoll.* **2014**, *39*, 301–318. [CrossRef]

251. Nishinari, K.; Fang, Y.; Nagano, T.; Guo, S.; Wang, R. *Soy as a Food Ingredient*, 2nd ed.; Yada, R.Y., Ed.; Elsevier Science and Technology: Cambridge, UK, 2018; ISBN 9780081007297.

252. Singh, P.; Kumar, R.; Sabapathy, S.N.; Bawa, A.S. Functional and edible uses of soy protein products. *Compr. Rev. Food Sci. Food Saf.* **2008**, *7*, 14–28. [CrossRef]
253. Saio, K.; Kamiya, M.; Watanabe, T.; Koyama, E.; Yamazaki, S.; Chem, A.B.; Wolf, W.J.; Babcock, T.E.; Smith, A.K.; Teramachi, Y.; et al. Food processing characteristics of soybean 11s and 7s proteins: Part I. Effect of difference of protein components among soybean varieties on formation of tofu-gel. *Agric. Biol. Chem.* **1969**, *33*, 1301–1308. [CrossRef]
254. Poysa, V.; Woodrow, L.; Yu, K. Effect of soy protein subunit composition on tofu quality. *Food Res. Int.* **2006**, *39*, 309–317. [CrossRef]
255. Zarkadas, C.G.; Gagnon, C.; Poysa, V.; Khanizadeh, S.; Cober, E.R.; Chang, V.; Gleddie, S. Protein quality and identification of the storage protein subunits of tofu and null soybean genotypes, using amino acid analysis, one- and two-dimensional gel electrophoresis, and tandem mass spectrometry. *Food Res. Int.* **2007**, *40*, 111–128. [CrossRef]
256. Takahashi, M.; Uematsu, Y.; Kashiwaba, K.; Yagasaki, K.; Hajika, M.; Matsunaga, R.; Komatsu, K.; Ishimoto, M. Accumulation of high levels of free amino acids in soybean seeds through integration of mutations conferring seed protein deficiency. *Planta* **2003**, *217*, 577–586. [CrossRef] [PubMed]
257. Kim, W.S.; Krishnan, H.B. Expression of an 11 kDa methionine-rich delta-zein in transgenic soybean results in the formation of two types of novel protein bodies in transitional cells situated between the vascular tissue and storage parenchyma cells. *Plant Biotechnol. J.* **2004**, *2*, 199–210. [CrossRef]
258. Kim, W.-S.; Jez, J.M.; Krishnan, H.B. Effects of proteome rebalancing and sulfur nutrition on the accumulation of methionine rich δ-zein in transgenic soybeans. *Front. Plant Sci.* **2014**, *5*, 1–12. [CrossRef]
259. Schmidt, M.A.; Herman, E.M. Proteome rebalancing in soybean seeds can be exploited to enhance foreign protein accumulation. *Plant Biotechnol. J.* **2008**, *6*, 832–842. [CrossRef]
260. Wieser, H. Chemistry of gluten proteins. *Food Microbiol.* **2007**, *24*, 115–119. [CrossRef] [PubMed]
261. Altenbach, S.B.; Tanaka, C.K.; Pineau, F.; Lupi, R.; Drouet, M.; Beaudouin, E.; Morisset, M.; Denery-Papini, S. Assessment of the allergenic potential of transgenic wheat (*Triticum aestivum*) with reduced levels of ω5-gliadins, the major sensitizing allergen in wheat-dependent exercise-induced anaphylaxis. *J. Agric. Food Chem.* **2015**, *63*, 9323–9332. [CrossRef] [PubMed]
262. Gil-Humanes, J.; Piston, F.; Tollefsen, S.; Sollid, L.M.; Barro, F. Effective shutdown in the expression of celiac disease-related wheat gliadin T-cell epitopes by RNA interference. *Proc. Natl. Acad. Sci. USA* **2010**, *107*, 17023–17028. [CrossRef] [PubMed]
263. Barro, F.; Rooke, L.; Bekes, F.; Gras, P.; Tatham, A.S.; Fido, R.; Lazzeri, P.A.; Shrewy, P.R.; Barcelo, P.; Tatha, A.S.; et al. Transformation of wheat with high molecular weight subunit genes results in improved functional properties. *Nat. Biotechnol.* **1997**, *15*, 871–875. [CrossRef] [PubMed]
264. Altpeter, F.; Vasil, V.; Srivastava, V.; Vasil, I.K. Integration and expression of the high-molecular-weight glutenin subunit 1Ax1 gene in wheat. *Nat. Biotechnol.* **1996**, *14*, 1246–1251. [CrossRef] [PubMed]
265. Blechl, A.E.; Anderson, O.D. Expression of a novel high-molecular-weight glutenin subunit gene in transgenic wheat. *Nat. Biotechnol.* **1996**, *14*, 875–879. [CrossRef]
266. Altenbach, S.B.; Tanaka, C.K.; Seabourn, B.W. Silencing of omega-5 gliadins in transgenic wheat eliminates a major source of environmental variability and improves dough mixing properties of flour. *BMC Plant Biol.* **2014**, *14*. [CrossRef]
267. Pistón, F.; Gil-Humanes, J.; Rodríguez-Quijano, M.; Barro, F. Down-Regulating γ-Gliadins in bread wheat leads to Non-Specific increases in other gluten proteins and has no major effect on dough gluten strength. *PLoS ONE* **2011**, *6*, 1–10. [CrossRef]
268. Troise, A.D.; Wilkin, J.D.; Fiore, A. Impact of rapeseed press-cake on Maillard reaction in a cookie model system. *Food Chem.* **2018**, *243*, 365–372. [CrossRef]
269. Altenbach, S.B.; Kuo, C.C.; Staraci, L.C.; Pearson, K.W.; Wainwright, C.; Georgescu, A.; Townsend, J. Accumulation of a Brazil nut albumin in seeds of transgenic canola results in enhanced levels of seed protein methionine. *Plant Mol. Biol.* **1992**, *18*, 235–245. [CrossRef]
270. Guerche, P.; De Almeida, E.R.P.; Schwarztein, M.A.; Gander, E.; Krebbers, E.; Pelletier, G. Expression of the 2S albumin from Bertholletia excelsa in Brassica napus. *MGG Mol. Gen. Genet.* **1990**, *221*, 306–314. [CrossRef]
271. Kohno-Murase, J.; Murase, M.; Ichikawa, H.; Imamura, J. Effects of an antisense napin gene on seed storage compounds in transgenic *Brassica napus* seeds. *Plant Mol. Biol.* **1994**, *26*, 1115–1124. [CrossRef] [PubMed]
272. Rolletschek, H.; Schwender, J.; König, C.; Chapman, K.D.; Romsdahl, T.; Lorenz, C.; Braun, H.P.; Denolf, P.; van Audenhove, K.; Munz, E.; et al. Cellular plasticity in response to suppression of storage proteins in the *Brassica napus* embryo. *Plant Cell* **2020**, *32*, 2383–2401. [CrossRef] [PubMed]
273. Edwards, D.; Batley, J.; Snowdon, R.J. Accessing complex crop genomes with next-generation sequencing. *Theor. Appl. Genet.* **2013**, *126*, 1–11. [CrossRef] [PubMed]
274. Zhu, C.; Gore, M.; Buckler, E.S.; Yu, J. Status and Prospects of Association Mapping in Plants. *Plant Genome* **2008**, *1*, 5–20. [CrossRef]
275. Mason, A.S.; Higgins, E.E.; Snowdon, R.J.; Batley, J.; Stein, A.; Werner, C.; Parkin, I.A.P. A user guide to the Brassica 60K Illumina Infinium^TM SNP genotyping array. *Theor. Appl. Genet.* **2017**, *130*, 621–633. [CrossRef]
276. Clarke, W.E.; Higgins, E.E.; Plieske, J.; Wieseke, R.; Sidebottom, C.; Khedikar, Y.; Batley, J.; Edwards, D.; Meng, J.; Li, R.; et al. A high-density SNP genotyping array for Brassica napus and its ancestral diploid species based on optimised selection of single-locus markers in the allotetraploid genome. *Theor. Appl. Genet.* **2016**, *129*, 1887–1899. [CrossRef]

277. You, Q.; Yang, X.; Peng, Z.; Xu, L.; Wang, J. Development and applications of a high throughput genotyping tool for polyploid crops: Single nucleotide polymorphism (SNP) array. *Front. Plant Sci.* **2018**, *9*. [CrossRef]

278. Baird, N.A.; Etter, P.D.; Atwood, T.S.; Currey, M.C.; Shiver, A.L.; Lewis, Z.A.; Selker, E.U.; Cresko, W.A.; Johnson, E.A. Rapid SNP discovery and genetic mapping using sequenced RAD markers. *PLoS ONE* **2008**, *3*, 1–7. [CrossRef]

279. Elshire, R.J.; Glaubitz, J.C.; Sun, Q.; Poland, J.A.; Kawamoto, K.; Buckler, E.S.; Mitchell, S.E. A robust, simple genotyping-by-sequencing (GBS) approach for high diversity species. *PLoS ONE* **2011**, *6*, 1–10. [CrossRef]

280. Scheben, A.; Batley, J.; Edwards, D. Genotyping-by-sequencing approaches to characterize crop genomes: Choosing the right tool for the right application. *Plant Biotechnol. J.* **2017**, *15*, 149–161. [CrossRef]

281. Myles, S.; Peiffer, J.; Brown, P.J.; Ersoz, E.S.; Zhang, Z.; Costich, D.E.; Buckler, E. Association mapping: Critical considerations shift from genotyping to experimental design. *Plant Cell* **2009**, *21*, 2194–2202. [CrossRef]

282. Xu, Y.; Li, P.; Yang, Z.; Xu, C. Genetic mapping of quantitative trait loci in crops. *Crop J.* **2017**, *5*, 175–184. [CrossRef]

283. Cockram, J.; Mackay, I. Genetic Mapping Populations for Conducting High-Resolution Trait Mapping in Plants. *Adv. Biochem. Eng. Biotechnol.* **2018**, *164*, 109–138.

284. Rasheed, A.; Hao, Y.; Xia, X.; Khan, A.; Xu, Y.; Varshney, R.K.; He, Z. Crop Breeding Chips and Genotyping Platforms: Progress, Challenges, and Perspectives. *Mol. Plant* **2017**, *10*, 1047–1064. [CrossRef]

285. Lorieux, M. MapDisto: Fast and efficient computation of genetic linkage maps. *Mol. Breed.* **2012**, *30*, 1231–1235. [CrossRef]

286. Meng, L.; Li, H.; Zhang, L.; Wang, J. QTL IciMapping: Integrated software for genetic linkage map construction and quantitative trait locus mapping in biparental populations. *Crop J.* **2015**, *3*, 269–283. [CrossRef]

287. Soto-Cerda, B.J. Association Mapping in Plant Genomes. In *Genetic Diversity in Plants*; InTech: London, UK, 2012.

288. Yu, J.; Buckler, E.S. Genetic association mapping and genome organization of maize. *Curr. Opin. Biotechnol.* **2006**, *17*, 155–160. [CrossRef]

289. Risch, N.; Merikangas, K. The future of genetic studies of complex human diseases. *Science* **1996**, *273*, 1516–1517. [CrossRef]

290. Thornsberry, J.M.; Goodman, M.M.; Doebley, J.; Kresovich, S.; Nielsen, D.; Buckler, E.S. Dwarf8 polymorphisms associate with variation in flowering time. *Nat. Genet.* **2001**, *28*, 286–289. [CrossRef] [PubMed]

291. Liu, H.J.; Yan, J. Crop genome-wide association study: A harvest of biological relevance. *Plant J.* **2019**, *97*, 8–18. [CrossRef]

292. Tian, D.; Wang, P.; Tang, B.; Teng, X.; Li, C.; Liu, X.; Zou, D.; Song, S.; Zhang, Z. GWAS Atlas: A curated resource of genome-wide variant-trait associations in plants and animals. *Nucleic Acids Res.* **2020**, *48*, D927–D932. [CrossRef] [PubMed]

293. Ersoz, E.S.; Yu, J.; Buckler, E.S. Applications of Linkage Disequilibrium and Association Mapping in Maize. In *Molecular Genetic Approaches to Maize Improvement*; Springer Berlin Heidelberg: Berlin, Heidelberg, 2009; pp. 173–195.

294. Li, J.; Bus, A.; Spamer, V.; Stich, B. Comparison of statistical models for nested association mapping in rapeseed (Brassica napus L.) through computer simulations. *BMC Plant Biol.* **2016**, *16*. [CrossRef] [PubMed]

295. Huang, X.; Han, B. Natural Variations and Genome-Wide Association Studies in Crop Plants. *Annu. Rev. Plant Biol.* **2014**, *65*, 531–551. [CrossRef] [PubMed]

296. Tang, Y.; Liu, X.; Wang, J.; Li, M.; Wang, Q.; Tian, F.; Su, Z.; Pan, Y.; Liu, D.; Lipka, A.E.; et al. GAPIT version 2: An enhanced integrated tool for genomic association and prediction. *Plant Genome* **2016**, *9*. [CrossRef]

297. Chen, P.; Shen, Z.; Ming, L.; Li, Y.; Dan, W.; Lou, G.; Peng, B.; Wu, B.; Li, Y.; Zhao, D.; et al. Genetic basis of variation in rice seed storage protein (Albumin, globulin, prolamin, and glutelin) content revealed by genome-wide association analysis. *Front. Plant Sci.* **2018**, *9*, 1–15. [CrossRef]

298. Deng, M.; Li, D.; Luo, J.; Xiao, Y.; Liu, H.; Pan, Q.; Zhang, X.; Jin, M.; Zhao, M.; Yan, J. The genetic architecture of amino acids dissection by association and linkage analysis in maize. *Plant Biotechnol. J.* **2017**, *15*, 1250–1263. [CrossRef] [PubMed]

299. Peng, Y.; Liu, H.; Chen, J.; Shi, T.; Zhang, C.; Sun, D.; He, Z.; Hao, Y.; Chen, W. Genome-wide association studies of free amino acid levels by six multi-locus models in bread wheat. *Front. Plant Sci.* **2018**, *9*, 1–9. [CrossRef]

300. Lee, S.; Van, K.; Sung, M.; Nelson, R.; LaMantia, J.; McHale, L.K.; Mian, M.A.R. Genome-wide association study of seed protein, oil and amino acid contents in soybean from maturity groups I to IV. *Theor. Appl. Genet.* **2019**, *132*, 1639–1659. [CrossRef]

301. Angelovici, R.; Lipka, A.E.; Deason, N.; Gonzalez-Jorge, S.; Lin, H.; Cepela, J.; Buell, R.; Gore, M.A.; DellaPenna, D. Genome-wide analysis of branched-chain amino acid levels in Arabidopsis Seeds. *Plant Cell* **2013**, *25*, 4827–4843. [CrossRef]

302. Hatzig, S.; Breuer, F.; Nesi, N.; Ducournau, S.; Wagner, M.H.; Leckband, G.; Abbadi, A.; Snowdon, R.J. Hidden effects of seed quality breeding on germination in oilseed rape (Brassica napus L.). *Front. Plant Sci.* **2018**, *9*, 1–13. [CrossRef]

303. Qian, L.; Qian, W.; Snowdon, R.J. Sub-genomic selection patterns as a signature of breeding in the allopolyploid Brassica napus genome. *BMC Genomics* **2014**, *15*, 1–17. [CrossRef]

304. Li, F.; Chen, B.; Xu, K.; Wu, J.; Song, W.; Bancroft, I.; Harper, A.L.; Trick, M.; Liu, S.; Gao, G.; et al. Genome-wide association study dissects the genetic architecture of seed weight and seed quality in rapeseed (Brassica napus L.). *DNA Res.* **2014**, *21*, 355–367. [CrossRef]

305. Qu, C.; Jia, L.; Fu, F.; Zhao, H.; Lu, K.; Wei, L.; Xu, X.; Liang, Y.; Li, S.; Wang, R.; et al. Genome-wide association mapping and Identification of candidate genes for fatty acid composition in Brassica napus L. using SNP markers. *BMC Genom.* **2017**, *18*, 1–17. [CrossRef]

306. Guan, M.; Huang, X.; Xiao, Z.; Jia, L.; Wang, S.; Zhu, M.; Qiao, C.; Wei, L.; Xu, X.; Liang, Y.; et al. Association mapping analysis of fatty acid content in different ecotypic rapeseed using mrMLM. *Front. Plant Sci.* **2019**, *9*. [CrossRef]

307. Wang, J.; Jian, H.; Wei, L.; Qu, C.; Xu, X.; Lu, K.; Qian, W.; Li, J.; Li, M.; Liu, L. Genome-Wide Analysis of Seed Acid Detergent Lignin (ADL) and hull content in rapeseed (Brassica napus L.). *PLoS ONE* **2015**, *10*, 1–18. [CrossRef] [PubMed]

308. Liu, S.; Huang, H.; Yi, X.; Zhang, Y.; Yang, Q.; Zhang, C.; Fan, C.; Zhou, Y. Dissection of genetic architecture for glucosinolate accumulations in leaves and seeds of Brassica napus by genome-wide association study. *Plant Biotechnol. J.* **2020**, *18*, 1472–1484. [CrossRef]

309. Qu, C.M.; Li, S.M.; Duan, X.J.; Fan, J.H.; Jia, L.D.; Zhao, H.Y.; Lu, K.; Li, J.N.; Xu, X.F.; Wang, R. Identification of candidate genes for seed glucosinolate content using association mapping in brassica napus L. *Genes* **2015**, *6*, 1215–1229. [CrossRef] [PubMed]

310. Jan, H.U.; Guan, M.; Yao, M.; Liu, W.; Wei, D.; Abbadi, A.; Zheng, M.; He, X.; Chen, H.; Guan, C.; et al. Genome-wide haplotype analysis improves trait predictions in Brassica napus hybrids. *Plant Sci.* **2019**, *283*, 157–164. [CrossRef]

311. Werner, C.R.; Voss-Fels, K.P.; Miller, C.N.; Qian, W.; Hua, W.; Guan, C.-Y.; Snowdon, R.J.; Qian, L. Effective Genomic Selection in a Narrow-Genepool Crop with Low-Density Markers: Asian Rapeseed as an Example. *Plant Genome* **2018**, *11*, 170084. [CrossRef]

312. Qian, L.; Qian, W.; Snowdon, R.J. Haplotype hitchhiking promotes trait coselection in Brassica napus. *Plant Biotechnol. J.* **2016**, *14*, 1578–1588. [CrossRef]

313. Liu, S.; Fan, C.; Li, J.; Cai, G.; Yang, Q.; Wu, J.; Yi, X.; Zhang, C.; Zhou, Y. A genome-wide association study reveals novel elite allelic variations in seed oil content of Brassica napus. *Theor. Appl. Genet.* **2016**, *129*, 1203–1215. [CrossRef] [PubMed]

314. Sun, F.; Liu, J.; Hua, W.; Sun, X.; Wang, X.; Wang, H. Identification of stable QTLs for seed oil content by combined linkage and association mapping in Brassica napus. *Plant Sci.* **2016**, *252*, 388–399. [CrossRef] [PubMed]

315. Qian, L.; Voss-Fels, K.; Cui, Y.; Jan, H.U.; Samans, B.; Obermeier, C.; Qian, W.; Snowdon, R.J. Deletion of a Stay-Green Gene Associates with Adaptive Selection in Brassica napus. *Mol. Plant* **2016**, *9*, 1559–1569. [CrossRef]

316. Lin, Y.; Pajak, A.; Marsolais, F.; McCourt, P.; Riggs, C.D. Characterization of a Cruciferin deficient mutant of Arabidopsis and its utility for overexpression of foreign proteins in plants. *PLoS ONE* **2013**, *8*, 1–8. [CrossRef]

317. Huang, S.; Adams, W.R.; Zhou, Q.; Malloy, K.P.; Voyles, D.A.; Anthony, J.; Kriz, A.L.; Luethy, M.H. Improving Nutritional Quality of Maize Proteins by Expressing Sense and Antisense Zein Genes. *J. Agric. Food Chem.* **2004**, *52*, 1958–1964. [CrossRef]

318. Lange, M.; Vincze, E.; Wieser, H.; Schjoerring, J.K.; Holm, P.B. Suppression of C-hordein synthesis in barley by antisense constructs results in a more balanced amino acid composition. *J. Agric. Food Chem.* **2007**, *55*, 6074–6081. [CrossRef]

319. Goossens, A.; Van Montagu, M.; Angenon, G. Co-introduction of an antisense gene for an endogenous seed storage protein can increase expression of a transgene in *Arabidopsis thaliana* seeds. *FEBS Lett.* **1999**, *456*, 160–164. [CrossRef]

320. Withana-Gamage, T.S.; Hegedus, D.D.; Qiu, X.; Yu, P.; May, T.; Lydiate, D.; Wanasundara, J.P.D. Characterization of *Arabidopsis thaliana* lines with altered seed storage protein profiles using synchrotron-powered FT-IR spectromicroscopy. *J. Agric. Food Chem.* **2013**, *61*, 901–912. [CrossRef]

321. Hegedus, D.D.; Baron, M.; Labbe, N.; Coutu, C.; Lydiate, D.; Lui, H.; Rozwadowski, K. A strategy for targeting recombinant proteins to protein storage vacuoles by fusion to *Brassica napus* napin in napin-depleted seeds. *Protein Expr. Purif.* **2014**, *95*, 162–168. [CrossRef]

322. Langner, T.; Kamoun, S.; Belhaj, K. CRISPR crops: Plant genome editing toward disease resistance. *Annu. Rev. Phytopathol.* **2018**, *56*, 479–512. [CrossRef] [PubMed]

323. Jinek, M.; Chylinski, K.; Fonfara, I.; Hauer, M.; Doudna, J.A.; Charpentier, E. A Programmable Dual-RNA—Guided DNA Endonuclease in Adaptive BActerial Immunity. *Science* **2012**, *337*, 816–822. [CrossRef] [PubMed]

324. Yang, H.; Wu, J.J.; Tang, T.; De Liu, K.; Dai, C. CRISPR/Cas9-mediated genome editing efficiently creates specific mutations at multiple loci using one sgRNA in *Brassica napus*. *Sci. Rep.* **2017**, *7*, 1–13. [CrossRef] [PubMed]

325. Braatz, J.; Harloff, H.-J.J.; Mascher, M.; Stein, N.; Himmelbach, A.; Jung, C. CRISPR-Cas9 targeted mutagenesis leads to simultaneous modification of different homoeologous gene copies in polyploid oilseed rape (*Brassica napus*). *Plant Physiol.* **2017**, *174*, 935–942. [CrossRef]

326. Sun, Q.; Lin, L.; Liu, D.; Wu, D.; Fang, Y.; Wu, J.; Wang, Y. CRISPR/Cas9-mediated multiplex genome editing of the *BnWRKY11* and *BnWRKY70* genes in Brassica napus L. *Int. J. Mol. Sci.* **2018**, *19*, 1–19. [CrossRef]

327. Okuzaki, A.; Ogawa, T.; Koizuka, C.; Kaneko, K.; Inaba, M.; Imamura, J.; Koizuka, N. CRISPR/Cas9-mediated genome editing of the fatty acid desaturase 2 gene in *Brassica napus*. *Plant Physiol. Biochem.* **2018**, *131*, 63–69. [CrossRef] [PubMed]

328. Li, C.; Hao, M.; Wang, W.; Wang, H.; Chen, F.; Chu, W.; Zhang, B.; Mei, D.; Cheng, H.; Hu, Q. An Efficient CRISPR/Cas9 Platform for Rapidly Generating Simultaneous Mutagenesis of Multiple Gene Homoeologs in Allotetraploid Oilseed Rape. *Front. Plant Sci.* **2018**, *9*, 1–9. [CrossRef] [PubMed]

329. Hu, L.; Zhang, H.; Yang, Q.; Meng, Q.; Han, S.; Nwafor, C.C.; Khan, M.H.U.; Fan, C.; Zhou, Y. Promoter variations in a homeobox gene, *BnA10.LMI1*, determine lobed leaves in rapeseed (*Brassica napus* L.). *Theor. Appl. Genet.* **2018**, *131*, 2699–2708. [CrossRef] [PubMed]

*Article*

# Investigation of Quinoa Seeds Fractions and Their Application in Wheat Bread Production

Ionica Coţovanu [1], Mădălina Ungureanu-Iuga [1,2,*] and Silvia Mironeasa [1,*]

1   Faculty of Food Engineering, Ştefan cel Mare University of Suceava, 13th Universităţii Street, 720229 Suceava, Romania; ionica.cotovanu@usm.ro
2   Integrated Center for Research, Development and Innovation in Advanced Materials, Nanotechnologies and Distributed Systems for Fabrication and Control (MANSiD), Ştefan cel Mare University of Suceava, 13th University Street, 720229 Suceava, Romania
*   Correspondence: madalina.iuga@usm.ro (M.U.-I.); silviam@fia.usv.ro (S.M.)



**Abstract:** The present study aimed to investigate the influence of quinoa fractions (QF) on the chemical components of wheat flour (WF), dough rheological properties, and baking performance of wheat bread. The microstructure and molecular conformations of QF fractions were dependent to the particle size. The protein, lipids, and ash contents of composite flours increased with the increase of QF addition level, while particle size (PS) decreased these parameters as follows: Medium > Small > Large, the values being higher compared with the control (WF). QF addition raised dough tenacity from 86.33 to 117.00 mm $H_2O$, except for the small fraction, and decreased the extensibility from 94.00 to 26.00 mm, while PS determined an irregular trend. The highest QF addition levels and PS led to the highest dough viscoelastic moduli (55,420 Pa for QL_20, 65245 Pa for QM_20 and 48305 Pa for QS_20, respectively). Gradual increase of QF determined dough hardness increase and adhesiveness decrease. Bread firmness, springiness, and gumminess rises were proportional to the addition level. The volume, elasticity, and porosity of bread decreased with QF addition. Flour and bread crust and crumb color parameters were also influenced by QF addition with different PS.

**Keywords:** quinoa seed fractions; particle size; wheat bread; addition level



**Citation:** Coţovanu, I.; Ungureanu-Iuga, M.; Mironeasa, S. Investigation of Quinoa Seeds Fractions and Their Application in Wheat Bread Production. *Plants* 2021, *10*, 2150. https://doi.org/10.3390/plants10102150

Academic Editor: Ivo Vaz de Oliveira

Received: 6 September 2021
Accepted: 7 October 2021
Published: 11 October 2021

**Publisher's Note:** MDPI stays neutral with regard to jurisdictional claims in published maps and institutional affiliations.

## 1. Introduction

Recently, there has been an increasing emphasis on a healthier lifestyle and healthy eating habits. Refined wheat flour has lower nutritional value: fewer fibers, vitamins, minerals, and phytochemicals than whole-grain wheat flour [1]. Bakery products from refined wheat flour are considered to be nutritionally poor, as the wheat proteins are deficient in essential amino acids such as lysine, tryptophan, and threonine [2,3]. The partial replacement of refined wheat flour with flours made from different crops rich in bioactive compounds has become a necessity during the last years.

Quinoa (*Chenopodium quinoa* Willd.) is an endemic grain that has attracted much attention in recent times due to its health and wellness benefits [4]. This species, is part of the *Chenopodiaceae* family and has been cultivated for centuries in the Andean countries of Peru and Bolivia [5]. Quinoa reveals a lack of gluten and plays a big role in the human diet because it covers half of people's daily energy and protein needs [6]. Quinoa is a complete food, being a source of proteins with high biological value, carbohydrates with a low glycemic index, high-quality oil, vitamins (thiamine, riboflavin, niacin, and vitamin E), minerals (magnesium, potassium, zinc, and manganese), and bioactive compounds (dietary fiber, phytosterols, polyphenols and flavonoids) [7–9]. Amino acids from quinoa are represented by lysine (twice than wheat), histidine, isoleucine, leucine, tryptophan, and aromatic amino acids (phenylalanine and tyrosine) [8,10]. Quinoa seeds have three main storage compartments (from center to edge): a large central perisperm surrounded by a peripheral embryo or germ, and an endosperm [11]. The starch granules are stored mainly in the perisperm which constitutes almost 40% of the seed mass, while protein, lipids, and minerals are found mostly in the embryo and endosperm [11,12]. Knowing the quinoa seeds morphology is very important for obtaining different enriched fractions because the quality of the baked products is interdependent on the constituents and the particle size of the flour used. Particle size modified the hydration properties of flour and influenced the dough's rheological properties [13].

The dynamic rheological testing methods indicate the viscoelastic behavior of the WF-QF dough given by the interactions between quinoa fractions and wheat gluten network. QF addition caused modifications of the rheological and textural parameters of wheat dough as a result of the gluten dilution [14–16]. The ingredients rich in fibers and starch could impact the gas retention of dough, leading to lower bread volume, porosity and elasticity, and higher firmness [17,18], the changes' magnitude depending on the addition level and particle size. Xiaoxuan et al. [19] demonstrated that the addition of whole quinoa to wheat bread resulted in a decrease of the final product specific volume, while the texture parameters in terms of hardness and chewiness where not significantly influenced at addition levels smaller than 20%. Another study conducted by Calderelli et al. [20] showed that wheat bread enriched with quinoa flour was acceptable from a sensory point of view and presented a high content of protein. Stikic et al. [21] reported positive effects of quinoa flours on the rheological characteristics of wheat dough, while bread-specific volume decreased slightly. According to the results obtained by El-Sohaimy et al. [22], the addition of quinoa in flat bread dough had a slight effect on the rheological characteristics, but did not determine dough deformation. Kurek and Sokolova [17] stated that wheat bread porosity increased due to the protein content of quinoa flour added. The same authors reported that the interaction between particle size and quinoa flour level showed a significant influence on wheat bread chemical and textural properties, particle size having a crucial effect on firmness parameter [17]. In the study of Xu et al. [23], it was demonstrated that the baking performance of wheat bread was not significantly affected by 5% quinoa addition level, while at levels higher than 10% smaller specific volume, increased hardness, and coarse porosity was observed due to the changes of gluten secondary structure and gluten dilution effect.

QF incorporation may impact dough rheology, which could provide information about its processability and hence could predict the baked good quality. The evaluation of wheat flour replacement with QF particle sizes at different addition levels can be useful for enhanced baked goods development, for choosing the appropriate recipes and manufacturing techniques. There are few studies on quinoa-wheat dough proprieties, but no previous studies concerning how the variation in particle size can influence the physico-chemical characteristics, complete texture profile, and dynamic rheological properties have been carried out until now. This work aimed to study the effect of wheat-quinoa composite flour, mix obtained with three different QF particle sizes and four addition levels in wheat flour on dough rheology and bread quality, in relation to the microstructure, molecular conformation, and physico-chemical characteristics of the raw materials.

## 2. Materials and Methods

### 2.1. Materials

The research was performed on wheat flour type 650 (WF) (harvest 2020) obtained from S.C. MOPAN S.A. (Suceava, Romania), which showed the following characteristics: 14.0% moisture, 12.60% protein, 1.40% fat, 0.65% ash, wet gluten 30%, gluten deformation index 6 mm, and falling number 312.0 s. White quinoa seeds were provided by the SANOVITA (ECUADOR) and were characterized by: moisture content 13.28%, fat content 5.61%, protein content 14.12%, and ash content 2.00%, reported to dried substances. Salt (S.C.SANOVITA S.R.L., Vâlcea, Romania) and fresh Saccharomyces cerevisiae yeast (S.C. ROMPAK S.R.L., Pașcani, România) were acquired from the local market.

### 2.2. Quinoa Fractions Preparation

The quinoa seeds were ground separately with Grain Mill grinder (KitchenAid, Whirlpool Corporation, Benton Harbor, MI, USA), then they were sifted 30 min at 70 Hz with a Retsch Vibratory Sieve Shaker AS 200 basic (Haan, Germany) in order to produce three fractions with different particle sizes (PS): large (L > 300 µm), medium (180 > M < 300 µm), and small (S < 180 µm).

### 2.3. Sample's Formulations

Each fraction of QF, large (L), medium (M), and small (S) at four addition levels (5%, 10%, 15%, and 20%) were mixed for half an hour in a Yucebas Y21 mixer (Izmir, Turkey), resulting in the following samples: QF_5L, QF_5M, QF_5S, QF_10L, QF_10M, QF_10S, QF_15L, QM_15M, QM_15S, QF_20L, QF_20M, and QF_20S. Wheat flour was considered as control.

### 2.4. Physico-Chemical Characterization of the Formulated Flours

The WF-QF flours were characterized in agreement to ICC methods: moisture content (110/1), protein content (105/2), fat content (105/1), ash content (104/1), and andcarbohydrate content which was calculated by difference, as % of dry matter.

Composite flour colors were analyzed in triplicate by using a CR-700 colorimeter (Konica Minolta, Tokyo, Japan). The flour color characteristics analyzed were: $L^*$—lightness/darkness (0: black and 100: white), and the chromatic components $a^*$—intensity of green ($-a^*$) or red ($+a^*$); and $b^*$—the intensity of blue ($-b^*$) or yellow ($+b^*$).

### 2.5. Dough and Bread Manufacturing

Composite flour or WF (0.3 kg), salt (1.8%), and yeast (3%) were used in the bread manufacturing process. Water absorption capacities of the flours were previously tested on the Mixolab device and used in dough preparation. The dough samples were prepared following the biphasic procedure by mixing water, yeast, and half the amount of composite flour for the sourdough development at 30 ± 2 °C and 85% relative humidity (RH) for 2 h in a leavening chamber (PL2008, Piron, Cadoneghe, Padova, Italy). The leavened sourdough, other half part of WF-QF flour, and salt were kneaded for 10 min with a Kitchen Aid mixer (Whirlpool Corporation, Benton Harbor, MI, USA) and leavened at 30 ± 2 °C and 85% relative humidity (RH) for another 60 min in the same leavening chamber [24]. When fermentation was finished, the dough was divided into 400 g pieces, molded by hand, and leavened in aluminum trays for another 60 min (30 ± 2 °C and 85% RH). The leavened dough was baked at 220 ± 5 °C for 25 min in an oven (Caboto PF8004D, Cadoneghe, Padova, Italy).

### 2.6. Evaluation of Flours Microstructure

The microstructures of WF and QF fractions were evaluated trough electronic scanning microscopy by using a VEGA II LSH device (Tescan, Brno, Czech Republic), at an acceleration tension of 30 kV. The samples were fixed on double-sided adhesive carbon bands and the images were collected at 2000×, 1000×, 500×, and 100× magnifications.

### 2.7. Flours ATR FT-IR Spectra Collection

The ATR FT-IR spectra of WF and QF fractions were collected in triplicate from 650 to 4000 cm−1 wavenumbers on a Thermo Scientific Nicolet iS20 (Waltham, MA, USA) device, at a resolution of 4 cm$^{-1}$ by 32 scans. The molecular characteristics were identified according to previous data from the literature [25–27], by using OMNIC software.

### 2.8. Empirical Dough Rheology and Texture Profile Analysis

The viscoelastic behavior of WF-QF flour was determined on a Chopin Alveograph NG (La Garenne Cedex, France) following the standard method SR EN ISO 27971:2009.

Each Alveograph curve was analyzed for the following parameters: P (dough resistance), L (dough extensibility), G (swelling index), W (baking strength), and P/L ratio [28].

Texture profile analysis (TPA) was performed on a TVT-6700 texture analyzer (Perten Instruments, Hägersten, Sweden), following the procedure of Mironeasa, Iuga, Zaharia, and Mironeasa [29] with slight modifications. A 3.5 cm stainless steel cylindrical probe was used in a twice-compression test to compress the 50 g of sample up to 50% of its depth, at a test speed of 5.0 mm/s, trigger force of 20 g, and the interval time between two compressions was 12 s. Hardness, adhesiveness, springiness, and cohesiveness were recorded. The measurements were carried out in triplicate.

### 2.9. Fundamental Dough Rheology

A preliminary stress sweep test was performed to identify the limits of the linear viscoelastic region (LVR) in the samples in which increasing strain was applied, from 0.00 to 100 Pa, at constant oscillation frequency of 1 Hz, according to some indications [30]. The dough samples prepared without yeasts, at optimum water absorption capacity, were placed in a measuring system of a HAAKE MARS 40 rheometer (Thermo-HAAKE, Karlsruhe, Germany) with a parallel plate-plates geometry and rested for 5 min prior to testing [31]. The excess dough was removed and a layer of vaseline was applied to the exposed edge to protect it from loss of moisture.

A frequency sweep test from 0.01 to 20 Hz at 10 Pa stress, in the LVR, was applied to determine dough storage ($G'$) and loss modulus ($G''$), at 20 °C.

A temperature sweep test was performed at a constant strain of 0.10% and a frequency of 1 Hz, the dough being heated from 20 to 100 °C at a rate of $4.0 \pm 0.1$ °C per min. The storage ($G'$) and loss modulus ($G''$) were recorded as a function of temperature by using Rheowin software. The maximum gelatinization temperature ($T_{max}$) was considered at the maximum $G'$ value.

### 2.10. Physical Properties of Bread

Bread physical properties were measured in triplicate two hours after baking, in agreement to the Romanian standard procedure SR 91:2007 in terms of volume, porosity, elasticity, and color. Loaf-specific volume ($cm^3$) was found by employing the seed displacement procedure. Porosity was calculated based on a sample cylinder volume (60 mm height and 45.50 mm diameter). Elasticity was determined on a crumb cylinder that was pressed for 1 min until half of its height, and then left to recover for 1 min [32].

Color analysis was determined after bread cutting in half and the crumb and the crust color were measured in triplicate by using a CR-700 colorimeter (Konica Minolta, Tokyo, Japan). The bread color characteristics analyzed were $L^*$, $a^*$, and $b^*$.

### 2.11. Bread Texture Parameters Determination

The bread was cut into slices of 50 mm thickness for the texture properties determination (in triplicate) by using a TVT-6700 texture analyzer (Perten Instruments, Hägersten, Sweden). A 2.5 cm cylindrical stainless-steel probe was used to compress the sample twice to a penetration distance of 20% of its depth, at a test speed of 1.0 mm/s, trigger force of 5 g, with an interval of 15 s between compressions. Firmness, springiness, gumminess, and cohesiveness were registered.

### 2.12. Statistical Analysis

Statistical software SPSS 25.0 (trial version) (IBM, New York, NY, USA) was used to calculate the means and standard deviations for all parameters. Statistically significant differences between parameters were determined by two-way analysis of variance with Tukey's test at $p \leq 0.05$ significance level. A principal component analysis (PCA) was applied to observe the relationships between the WF-QF flour chemical constituents, dough rheological measurements, and bread characteristics.

## 3. Results and Discussions

### *3.1. Microstructure of Flours*

The microstructure of QF fractions and WF at different magnifications is presented in Figure 1.

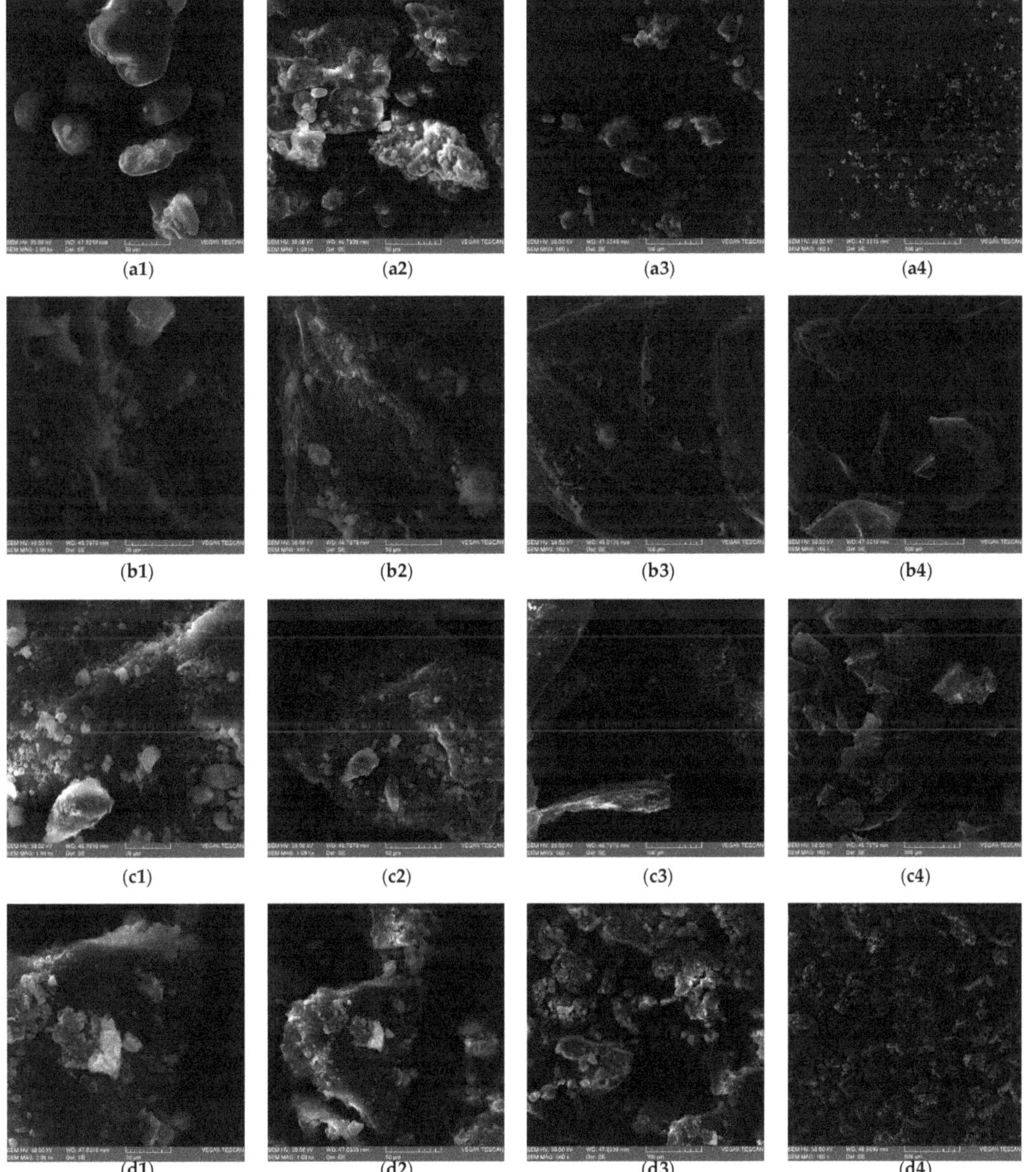

**Figure 1.** Microstructure of wheat flour (**a1–a4**) and quinoa flours fraction L (**b1–b4**), fraction M (**c1–c4**), and fraction S (**d1–d4**) at different magnifications.

WF structure was composed of starch grains surrounded by gluten proteins (Figure 1a). QF fractionation caused the decrease of PS, the particles presenting polygonal, angular or irregular shapes, similar to the results presented by Romano, Masi, Nicolai, Falciano and Ferrantia [33], and by Alvarez-Jubete, Auty, Arendt, and Gallagher [34]. A more uniform structure of quinoa flour was observed in S fraction (Figure 1(d1–d4)) compared with L (Figure 1(b1–b4)) and M (Figure 1(c1–c4)). Starch grains of rounded and lenticular shapes were present in L and M fractions, while in the case of S fraction irregular starch grains were observed, probably due to the damage caused during milling. Quinoa seeds fractionation led to changes in QF structure, depending on the particle size, which could explain the physico-chemical, rheological, and technological characteristics of flour, dough, and bread.

### 3.2. ATR FT-IR Spectra of Flours

The FT-IR spectra of WF and quinoa fractions are shown in Figure 2. There were differences in absorbances between L and M or S quinoa fractions, while the differences between S and M fractions were small. The intensities of the peaks increased as the PS decreased from L < M < S, while WF peaks were between L and M quinoa fractions.

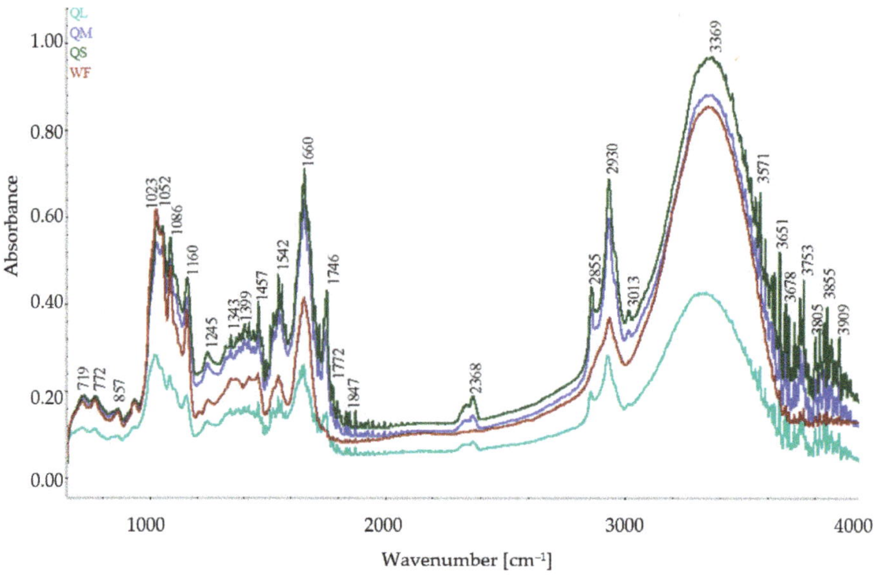



**Figure 2.** FT-IR spectra of wheat flour and quinoa flours fractions.

The starch structure identified at about 1100–900 cm$^{-1}$, amide I at 1600–1700 cm$^{-1}$, amide II at 1550 cm$^{-1}$, lipids at 1750 cm$^{-1}$, and 2800–3000 cm$^{-1}$ [26,35] were the most prominent peaks (Figure 2). The band at about 3300 cm$^{-1}$ is given by the stretching vibration of -OH, possibly due to the presence of water, galacturonic acid, arabinose, galactose, xylose, and glucose in quinoa fractions [25]. In the region of 900–1500 cm$^{-1}$, some signals were possibly given by the amylose-lipid complexes, amide III (at 1330–1230 cm$^{-1}$), or carbohydrates such as starch and cellulose [26,36]. The peak observed at 3369 cm$^{-1}$ could be attributed to the O-H stretching vibrations, the band found at 2855 cm$-1$ could be possibly assigned to the presence of CH2 and CH$_3$ groups from aldehydes/ketones [37], while the peak at 1746 cm$^{-1}$ could be attributed to the C=O carbonyl stretching [27]. The stretching given by alcohol and carbonyl groups identified could be possibly due to the chemical structure of quinoa saponins [27]. The band at 2930 cm$^{-1}$ could be given by the stretching vibrations of C-H groups which could be characteristic for polysaccharide-based polymers [25]. The peaks observed at 1660 and 1542 cm$^{-1}$ could give information about the protein amido acids and can reveal modifications in the secondary structure of proteins [37], while at 1086 cm$^{-1}$ possible information about pyranose structure of CH could be found [27]. The peak observed at 1021 cm$^{-1}$ could be attributed to the C-H bending from aromatic structures, similar results being obtained by Czekus et al. [27] for quinoa seeds. The bands present at 857, 772 and 719 cm$^{-1}$ could give information about the substitutions in aromatic rings characterized by aromatic C-H out-of-plane bend [27].

### 3.3. Physico-Chemical Properties of Composite Flours

Table 1 presents the effect of QF addition levels and PS on the composite flour's physicochemical properties. The studied factors had a significant ($p < 0.01$) effect on chemical parameters of composite flours. The moisture content decreased significantly ($p < 0.01$) with the rise of QF addition and PS decrease, due to the higher wear and possible heat generation that occurs during grinding of smaller PS flours, without prior process conditioning [38].

**Table 1.** Physico-chemical properties of composite flours as affected by quinoa flours fractions addition.

| Sample | Moisture (%) | Protein (%) | Lipids (%) | Ash (%) | Carbohydrates (%) | Color | | |
|---|---|---|---|---|---|---|---|---|
| | | | | | | $L^*$ | $a^*$ | $b^*$ |
| Control | 14.08 ± 0.08 [e] | 12.45 ± 0.15 [a] | 1.41 ± 0.01 [a] | 0.69 ± 0.04 [a] | 71.36 ± 0.01 [e] | 91.46 ± 0.15 [d] | -5.13 ± 0.03 [a] | 15.09 ± 0.07 [b] |
| QL_5 | 13.82 ±0.00 [dy] | 12.57 ± 0.03 [bx] | 1.65 ± 0.00 [by] | 0.70 ± 0.00 [bx] | 71.25 ± 0.04 [dz] | 90.87 ± 0.07 [cz] | −4.89 ± 0.02 [bx] | 14.98 ± 0.05 [abx] |
| QL_10 | 13.64 ± 0.00 [cy] | 12.54 ± 0.06 [cx] | 1.90 ± 0.00 [cy] | 0.76 ± 0.00 [cx] | 71.15 ± 0.06 [cz] | 90.43 ± 0.12 [bz] | −4.83 ± 0.01 [cx] | 14.51 ± 0.04 [ax] |
| QL_15 | 13.47 ± 0.01 [by] | 12.50 ± 0.09 [dx] | 2.16 ± 0.00 [dy] | 0.81 ± 0.00 [dx] | 71.05 ± 0.11 [bz] | 90.82 ± 0.25 [bz] | −4.99 ± 0.07 [cx] | 14.50 ± 0.11 [abx] |
| QL_20 | 13.29 ± 0.01 [ay] | 12.47 ± 0.12 [ex] | 2.41 ± 0.00 [ey] | 0.87 ± 0.00 [ex] | 70.95 ± 0.13 [az] | 89.38 ± 0.04 [az] | −4.60 ± 0.00 [dx] | 14.76 ± 0.02 [abx] |
| QM_5 | 13.80 ± 0.00 [dxy] | 12.91 ± 0.00 [bz] | 1.65 ± 0.00 [bxy] | 0.78 ± 0.00 [bz] | 70.85 ± 0.00 [dx] | 89.35 ± 0.08 [cx] | −4.85 ± 0.07 [bx] | 14.81 ± 0.13 [aby] |
| QM_10 | 13.60 ± 0.00 [cxy] | 13.22 ± 0.00 [cz] | 1.90 ± 0.00 [cxy] | 0.90 ± 0.00 [cz] | 70.36 ± 0.00 [cx] | 88.64 ± 0.11 [bx] | −4.72 ± 0.06 [cx] | 14.82 ± 0.12 [ay] |
| QM_15 | 13.47 ± 0.01 [bxy] | 13.54 ± 0.01 [dz] | 2.15 ± 0.00 [dxy] | 1.03 ± 0.00 [dz] | 69.87 ± 0.01 [bx] | 88.10 ± 0.12 [bx] | −4.69 ± 0.03 [cx] | 15.43 ± 0.52 [aby] |
| QM_20 | 13.21 ± 0.00 [axy] | 13.85 ± 0.03 [ez] | 2.40 ± 0.00 [exy] | 1.16 ± 0.01 [ez] | 69.38 ± 0.02 [ax] | 87.59 ± 0.23 [ax] | −4.53 ± 0.02 [dx] | 15.33 ± 0.19 [aby] |
| QS_5 | 13.79 ± 0.00 [dx] | 12.75 ± 0.00 [by] | 1.65 ± 0.00 [bx] | 0.74 ± 0.00 [by] | 71.06 ± 0.02 [dy] | 89.80 ± 0.23 [cy] | −4.93 ± 0.10 [by] | 14.91 ± 0.07 [aby] |
| QS_10 | 13.57 ± 0.03 [cx] | 12.91 ± 0.01 [cy] | 1.89 ± 0.00 [cx] | 0.84 ± 0.00 [cy] | 70.78 ± 0.04 [cy] | 89.05 ± 0.19 [by] | −4.80 ± 0.02 [cy] | 14.98 ± 0.04 [ay] |
| QS_15 | 13.36 ± 0.05 [bx] | 13.06 ± 0.02 [dy] | 2.14 ± 0.00 [dx] | 0.93 ± 0.00 [dy] | 70.50 ± 0.07 [by] | 88.71 ± 0.13 [by] | −4.69 ± 0.03 [cy] | 15.22 ± 0.08 [aby] |
| QS_20 | 13.15 ± 0.07 [ax] | 13.22 ± 0.03 [ey] | 2.39 ± 0.00 [ex] | 1.03 ± 0.00 [ey] | 70.20 ± 0.09 [ay] | 88.63 ± 0.05 [ay] | −4.59 ± 0.05 [dy] | 15.01 ± 0.04 [aby] |
| | | | | Two-way ANOVA *p* value | | | | |
| F1: | $p < 0.01$ | $p < 0.01$ | $p < 0.01$ | $p < 0.01$ | $p < 0.01$ | $p < 0.01$ | $p < 0.01$ | $p = 0.01$ |
| F2 | $p < 0.01$ | $p < 0.01$ | $p < 0.01$ | $p < 0.01$ | $p < 0.01$ | $p < 0.01$ | $p < 0.01$ | $p < 0.01$ |
| F1 × F2 | $p = 0.35$ | $p < 0.01$ | $p = 0.01$ | $p < 0.01$ | $p < 0.01$ | $p < 0.01$ | $p < 0.01$ | $p < 0.01$ |

F1: level of QF addition; F2: type of particle size; Mean values in the same column with different superscript letters indicates significantly difference ($p < 0.05$): [a–e] for QF addition level (0–20%); [x–z] for QF PS (L, M, and S). $L^*$-lightness; $a^*$-greenness; $b^*$-yelowness.

The protein content of composite flours was significantly influenced by QF addition level and increased when the QF addition increased in comparison with the control, due to a higher protein content of quinoa flour than wheat flour [39]. It was noticed that composite flours with medium particles had the highest protein content, followed by flours which contain small particles of QF, while the lower content of protein was found in flours with a large fraction of QF, probably as a result of the botanical structure of quinoa seeds, where proteins and minerals are localized mostly in the embryo and endosperm [12,40] and some part of the grain, richer in proteins, was broken in the form of small particles [41]. Additionally, these variations of protein content in flours enriched with quinoa flour fractions can be explained by the milling equipment used for grinding which differently influenced the structure of the endosperm (hard/soft) and type of endosperm cells (peripheral, prismatic, or central) [15]. Others authors that grounded quinoa seeds with coffee grinder found a higher protein content in small fraction [42]. The lipid content of the blended flours was significantly ($p < 0.01$) affected by the QF addition level and PS compared with wheat flour. The fat content of WF-QF composite flours increased gradually when QF addition level and PS increased, which could be explained by the lipid's localization in the cells of the endosperm and embryo [11]. The ash content of the wheat-quinoa formulated flours was significantly ($p < 0.01$) affected by the level and PS of QF, and increased with QF addition increase. The variations in ash content could be explained by the cell-wall material from the broken endosperm. Carbohydrate contents significantly ($p < 0.01$) decreased with the rise of QF addition and PS decrease. Similar findings regarding carbohydrates from WF-QF composite flours were found by ElSohaimy et al. [18]. Similar trends of the chemical compositions were previously reported by Coțovanu, Stoenescu and Mironeasa [39], by Ahmed, Thomas and Arfat [42], and Solaesa, Villanueva, Vela, and Ronda [15].

The color parameters, brightness, yellowness, and redness, were significantly ($p < 0.01$) influenced by the QF amount and PS. The lightness $L^*$ values decreased in all composite flours when the level of QF increased. The darker composite flour was observed at that was blended with QF medium PS, while the largest PS gave higher lightness of composite flours.

The redness ($a^*$) values significantly ($p < 0.01$) increased with the increase of QF quantity and with PS decrease, which indicates that the flour fraction turned more yellow and whitish and less red. This phenomenon could be explained by the increase of the particle surface area. The yellowness ($b^*$) values decreased when QF addition raised and increased with reducing PS. The yellowness could be explained by the carotenoid pigments [42]. Similar results were found for wheat-quinoa flour blends by Ahmed, Thomas and Arfat [42], and by Demir [43].

### 3.4. Dough Rheological Properties
### 3.4.1. Alveographic Parameters

The replacement of wheat flour, which contains gluten, is a major technological challenge because gluten is an essential structure-building protein in flour, responsible for the elastic and extensible properties needed to produce good quality bread [44]. The addition level and PS of QF had a significant ($p < 0.01$) effect on the dough's alveographic properties (Table 2).

**Table 2.** Alveographic parameters as affected by quinoa flours fractions.

| Sample | P (mm $H_2O$) | L (mm) | G | W ($\times 10^{-4}$ J) | P/L |
|---|---|---|---|---|---|
| Control | 86.33 ± 0.57 [a] | 94.00 ± 3.00 [d] | 21.55 ± 0.35 [d] | 253.00 ± 4.00 [d] | 0.92 ± 0.03 [a] |
| QL_5 | 88.50 ± 0.50 [by] | 46.50 ± 0.50 [cy] | 15.25 ± 0.05 [cy] | 167.50 ± 2.50 [cx] | 1.80 ± 0.06 [by] |
| QL_10 | 103.50 ± 0.50 [cy] | 39.50 ± 0.50 [by] | 13.80 ± 0.10 [by] | 166.00 ± 0.00 [cx] | 2.65 ± 0.01 [cy] |
| QL_15 | 104.00 ± 1.00 [cy] | 35.50 ± 0.50 [ay] | 12.75 ± 0.05 [ay] | 158.00 ± 2.00 [bx] | 3.39 ± 0.00 [ey] |
| QL_20 | 113.00 ± 1.00 [dy] | 32.00 ± 3.00 [ay] | 12.20 ± 0.60 [ay] | 142.00 ± 5.00 [ax] | 3.21 ± 0.01 [dy] |
| QM_5 | 102.50 ± 1.50 [bz] | 42.00 ± 1.00 [cx] | 14.75 ± 0.15 [cx] | 173.50 ± 0.50 [cx] | 2.44 ± 0.09 [bz] |
| QM_10 | 101.00 ± 0.00 [cz] | 38.00 ± 1.00 [bx] | 14.60 ± 0.00 [bx] | 172.00 ± 1.73 [cx] | 2.35 ± 0.00 [cz] |
| QM_15 | 113.00 ± 3.00 [cz] | 29.00 ± 0.00 [ax] | 12.00 ± 0.00 [ax] | 138.50 ± 4.50 [bx] | 4.53 ± 0.11 [ez] |
| QM_20 | 117.00 ±3.00 [dz] | 26.00 ± 2.00 [ax] | 11.40 ± 0.40 [ax] | 141.50 ± 3.50 [ax] | 3.92 ± 0.37 [dz] |
| QS_5 | 98.50 ± 0.50 [bx] | 58.50 ± 1.50 [cz] | 17.00 ± 0.20 [cz] | 211.50 ± 8.31 [cx] | 1.66 ± 0.01 [bx] |
| QS_10 | 96.00 ± 1.00 [cx] | 53.00 ± 3.00 [bz] | 16.05 ± 0.35 [bz] | 179.50 ± 0.50 [cx] | 1.87 ± 0.03 [cx] |
| QS_15 | 92.00 ± 1.00 [cx] | 43.50 ± 1.50 [az] | 14.60 ± 0.20 [az] | 158.00 ± 2.00 [bx] | 2.22 ± 0.01 [ex] |
| QS_20 | 80.50 ± 1.50 [dx] | 37.00 ± 1.00 [az] | 13.55 ± 0.15 [az] | 117.50 ± 6.50 [ax] | 2.18 ± 0.01 [dx] |
| Two-way ANOVA $p$ value | | | | | |
| F1 | $p < 0.01$ | $p < 0.01$ | $p < 0.01$ | $p < 0.01$ | $p < 0.01$ |
| F2 | $p < 0.01$ | $p < 0.01$ | $p < 0.01$ | $p < 0.01$ | $p < 0.01$ |
| F1 × F2 | $p < 0.01$ | $p < 0.01$ | $p < 0.01$ | $p < 0.01$ | $p < 0.01$ |

F1: level of QF addition; F2: type of particle size; means in the same column with different superscripts letters indicate significant difference ($p < 0.01$): [a–e] for QF addition level (0–20%); and [x–z] for QF PS (L, M, and S). P—dough tenacity; L—dough extensibility; G—index of swelling; W—dough strength; P/L—curve configuration ratio.

QF-WF dough tenacity was statistically ($p < 0.01$) influenced by the addition level, type of QF PS, and the interaction between them. The increment of QF increased gradually with dough tenacity (P) with large and medium PS, while in small PS a decrease was observed. The highest dough tenacity was observed at sample QM_20 and could be possibly explained by the chemical composition of the fraction added. The interactions between the polysaccharides of the fiber and the wheat proteins could be responsible for these increments [45]. The addition of QF in wheat flour dough increased fat and protein content (Table 1), which has an opposite effect; P and W. Sluimer [45] indicated that dough with a low content of lipids is somewhat more flexible, with better machinability, while a high quantity of lipids causes the opposite effect. It can be observed that P increased when wheat flour content decreased, a phenomenon that can be explained by the dough gluten dilution, finding that is consistent with other works [14,46]. PS determined the increase of dough tenacity as follows: S. > L. > M. Dough extensibility index (L) was significantly ($p < 0.01$) affected by the addition level, PS, and their interaction.

A decrement in dough extensibility compared with the control dough was observed with the increased of QF addition level, probably because a preferential pathway for water absorption was created. Large and medium PS determined a decrease of extensibility with size reduction, while the dough with small PS had higher extensibility. The influence of PS on dough extensibility could be correlated with protein content from these PS, which are characterized by minor gluten formation, results which are in line with earlier reports [14]. The index of swelling (G) was significantly ($p < 0.01$) influenced by the factors in the same way as dough extensibility. Dough strength (W) was significantly ($p < 0.01$) affected by both factors, and by the interaction between them. A decrement in the dough strength was observed when QF addition level increased, while an increment was obtained with QF PS reduction, probably due to the addition of a non-gluten flour, which will lead to a gluten dilution and a decrease of its quality and quantity, similar to the results reported by Coțovanu and Mironeasa [24]. As the PS and levels of quinoa flour increased, P increased and L decreased, which resulted in an increase of the P/L ratio from 0.92 to 4.53.

### 3.4.2. Dynamic Rheological Parameters

Viscoelastic properties of dough play a key role in final products quality and the frequency sweep tests demonstrated that $G'$ and $G''$ values were significantly ($p < 0.01$) influenced by the QF addition level, type of PS, and by their interaction. $G'$ values were greater than $G''$ values (Figure S1 presented in the Supplementary Materials), so it can be stated that the dough had a viscoelastic behavior. $G'$ and $G''$ increased when QF addition level and PS increased. The highest $G'$ and $G''$ were observed at large PS which can be explained by the synergistic effect between starch amount and the large PS, the results being in line with those found by Solaesa et al. [15]. Significant differences ($p < 0.01$) between QF dough samples and control were observed in loss tangent (tan $\delta$) regarding QF addition level, although the samples with 5% and 10% addition levels did not show significant differences in this parameter. The decrease of tan $\delta$ was proportional with the QF increase for all the tested samples. Significant differences ($p < 0.01$) between the higher PS (L and M) and S were obtained. These variations in rheological properties of the gels could be explained by their different chemical composition (protein, lipid, carbohydrates) and molecular structures (Figure 2), shape, and size of starch granules (Figure 1).

The influence of QF addition level and PS on maximum gelatinization temperature ($T_{max}$) during heating WF-QF composite flour is presented in Table 3. It can be observed a significant ($p < 0.05$) increase of $T_{max}$ with quinoa addition level increase (Figure S2 from the Supplementary Materials), while only L and S fractions significantly affected ($p < 0.01$) this parameter.

An increase in $T_{max}$ values was observed with PS decrease, which may be explained by the high proteins and lipids content, and their low carbohydrates content (which is associated with starch) (Table 1), similar data being reported by Ahmed, Thomas and Arfat [42].

**Table 3.** Dynamic moduli and gelatinization temperatures as affected by quinoa flours fractions.

| Type of Sample | G′ at 1 Hz (Pa) | G″ at 1 Hz (Pa) | Tan δ at 1 Hz (adim.) | $T_{max}$ (°C) |
|---|---|---|---|---|
| Control | 26,370 ± 70.15 [a] | 9488 ± 60.00 [a] | 0.3598 ± 0.00 [d] | 82.74 ± 0.49 [c] |
| QL_5 | 44,600 ± 270.00 [by] | 17,125 ± 145.00 [cy] | 0.3839 ± 0.00 [cy] | 78.32 ± 0.05 [abx] |
| QL_10 | 47,150 ± 190.00 [cy] | 15,615 ± 25.00 [by] | 0.3311 ± 0.01 [by] | 78.87 ± 0.02 [bx] |
| QL_15 | 52,790 ± 285.00 [dy] | 19,970 ± 100.00 [dy] | 0.3782 ± 0.00 [dy] | 78.68 ± 0.19 [abx] |
| QL_20 | 55,420 ± 40.00 [ey] | 20,525 ± 535.00 [dy] | 0.3703 ± 0.00 [dy] | 78.97 ± 1.01 [ax] |
| QM_5 | 34,865 ± 525.00 [bz] | 11,240 ± 60.00 [cz] | 0.3223 ± 0.00 [cy] | 78.47 ± 0.05 [abxy] |
| QM_10 | 47,905 ± 615.00 [cz] | 16,935 ± 145.00 [bz] | 0.3535 ± 0.00 [by] | 79.06 ± 0.14 [bxy] |
| QM_15 | 57,440 ± 310.00 [dz] | 18,640 ± 170.00 [dz] | 0.3245 ± 0.00 [dy] | 79.45 ± 0.08 [abxy] |
| QM_20 | 65,245 ± 205.00 [ez] | 19,745 ± 95.00 [dz] | 0.3026 ± 0.00 [dy] | 79.41 ± 0.03 [axy] |
| QS_5 | 31,320 ± 280.00 [bx] | 10,175 ± 335.00 [cx] | 0.3248 ± 0.00 [cx] | 80.28 ± 0.15 [aby] |
| QS_10 | 32,360 ± 200.00 [cx] | 10,853 ± 116.50 [bx] | 0.3353 ± 0.00 [bx] | 80.45 ± 0.18 [by] |
| QS_15 | 39,260 ± 585.00 [dx] | 14,360 ± 420.00 [dx] | 0.3657 ± 0.00 [dx] | 78.74 ± 0.07 [aby] |
| QS_20 | 48,305 ± 240.00 [ex] | 15,725 ± 45.00 [dx] | 0.3255 ± 0.00 [dx] | 78.97 ± 0.11 [ay] |
| **Two-way ANOVA *p* value** | | | | |
| F1 | *p* < 0.01 | *p* < 0.01 | *p* < 0.01 | *p* < 0.01 |
| F2 | *p* < 0.01 | *p* < 0.01 | *p* < 0.01 | *p* < 0.01 |
| F1 × F2 | *p* < 0.01 | *p* < 0.01 | *p* < 0.01 | *p* < 0.01 |

F1: level of QF addition; F2: type of particle size; means in the same column with different superscript letters indicate significant difference (*p* < 0.01): [a–e] for QF addition level (0–20%); and [x–z] for QF PS (L, M, and S). G′—elastic modulus; G″—viscous modulus; tan δ—loss tangent; $T_{max}$—maximum gelatinization temperature.

### 3.4.3. Dough Texture Profile Analysis

The effect of QF addition level and PS on dough texture parameters is shown in Figure 3. Hardness increased with the addition level increase and PS decrease, probably due to 11S-type globulin and 2S albumins, which bound to each other through disulfide bridges and retain more water, than prolamins from wheat flour. This lack of gluten from the dough matrix leads to low gas retention, which forms a harder dough [16].

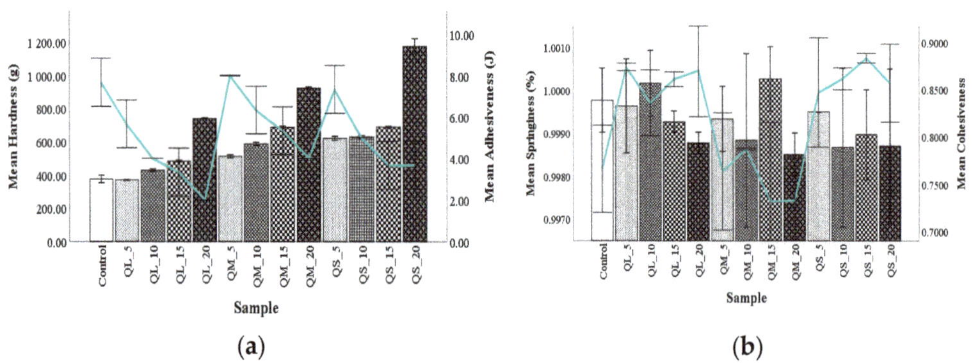

**(a)**

**(b)**

**Figure 3.** The effect of quinoa flours fractions on dough texture: hardness and adhesiveness (**a**), springiness and cohesiveness (**b**).

Adhesiveness decreased when the QF addition level increased, which can be explained by the gluten dilution, because it is well known that gliadin has a positive impact on the adhesiveness of the dough [29]. Springiness presented a decrease in comparison with WF dough, but samples QL_10 and QM_15 presented higher values due to the presence of prolamins from wheat flour (40–50%), which can make the wheat flour dough a little bit inelastic, which resulted in the springiness of dough with quinoa fractions that are richer in albumins [22]. Dough cohesiveness values decreased especially in samples formulated with medium PS, which were lower than the control sample and may be explained by the high lipid content from these blends, while for the other samples irregular trends were observed (Figure 3). Wheat gliadins presented a low resistance to extension, which could be responsible for the cohesion of the dough, and wheat glutenins for the dough's resistance [47].

### 3.5. Physical Properties of Bread

Quinoa flour addition level, its different PS, and the interaction between them significantly ($p < 0.05$) influenced bread characteristics. It is well known that gluten-free grains affect dough gas holding properties which will be negatively reflected in bread volume. QF addition level decreased the volume of bread from 378.13 to 260.00 cm$^3$ and from 2.45 to 1.81 cm$^3$/g respectively (Table 4), which could be explained by the gluten dilution of doughs with a higher amount of non-gluten flour. Final product quality is strongly influenced by the constituents of the ingredient added. Small fractions have a higher water absorption capacity, but resulted in low bread volume probably due to the intrinsic factors that affect the water-binding properties of flours with a relatively high protein content, factors that refer to amino acid composition, protein conformation, and surface polarity [48].

**Table 4.** Physical characteristics of bread as affected by quinoa flours fractions.

| Sample | Loaf Volume (cm$^3$) | Specific Volume (g/cm$^3$) | Porosity (%) | Elasticity (%) |
|---|---|---|---|---|
| Control | 378.70 ± 1.12 [e] | 2.45 ± 0.00 [e] | 64.33 ± 0.11 [b] | 91.72 ± 0.07 [b] |
| QL_5 | 372.60 ± 0.52 [dx] | 2.25 ± 0.02 [dxy] | 72.38 ± 0.16 [ex] | 97.92 ± 0.37 [ez] |
| QL_10 | 358.87 ± 1.02 [cx] | 2.20 ± 0.00 [cxy] | 67.93 ± 0.05 [dx] | 94.11 ± 0.84 [dz] |
| QL_15 | 335.27 ± 0.37 [bx] | 2.00 ± 0.06 [bxy] | 66.35 ± 0.34 [cx] | 93.17 ± 0.45 [cz] |
| QL_20 | 317.01 ± 1.24 [ax] | 1.93 ± 0.01 [axy] | 57.27 ± 0.52 [ax] | 89.99 ± 1.66 [az] |
| QM_5 | 371.30 ± 1.21 [dx] | 2.24 ± 0.01 [dy] | 72.47 ± 0.07 [ez] | 96.36 ± 0.29 [eyz] |
| QM_10 | 363.53 ± 1.27 [cx] | 2.22 ± 0.01 [cy] | 70.87 ± 0.46 [dz] | 94.51 ± 0.31 [dyz] |
| QM_15 | 338.86 ± 0.15 [bx] | 2.05 ± 0.00 [by] | 67.63 ± 0.81 [cz] | 93.48 ± 0.15 [cyz] |
| QM_20 | 318.63 ± 0.81 [ax] | 1.93 ± 0.00 [ay] | 66.32 ± 0.58 [az] | 89.74 ± 0.50 [ayz] |
| QS_5 | 356.66 ± 1.52 [dy] | 2.21 ± 0.02 [dx] | 71.97 ± 0.52 [ey] | 96.17 ± 0.10 [exy] |
| QS_10 | 347.33 ± 2.08 [cy] | 2.18 ± 0.00 [cx] | 70.51 ± 0.09 [dy] | 94.86 ± 0.93 [dxy] |
| QS_15 | 303.66 ± 3.51 [by] | 2.04 ± 0.03 [bx] | 66.63 ± 0.80 [cy] | 92.00 ± 0.63 [cxy] |
| QS_20 | 260.00 ± 3.00 [ay] | 1.81 ± 0.08 [ax] | 61.60 ± 1.01 [ay] | 87.72 ± 0.96 [axy] |
| Two-way ANOVA $p$ value | | | | |
| F1 | $p < 0.01$ | $p < 0.01$ | $p < 0.01$ | $p < 0.01$ |
| F2 | $p < 0.01$ | $p < 0.01$ | $p < 0.01$ | $p < 0.01$ |
| F1 × F2 | $p < 0.01$ | $p = 0.01$ | $p < 0.01$ | $p = 0.01$ |

F1: level of QF addition; F2: type of particle size; means in the same column with different superscript letters indicate significant difference ($p < 0.01$): [a]–[e] for QF addition level (0–20%); and [x]–[z] for QF PS (L, M, and S).

The addition of QF in WF had a significant effect on decreasing the dough strength. Although the WF dough strength was higher, because of its extensibility, the ability to preserve or increase the dough strength absorbed by the dough with the addition of quinoa flours was extremely low [49]. The competition between dietary fiber and starch for water leads to a limited starch swelling and gelatinization, which might be required to reduce the final gas volume fraction in the crumb [50]. Also, the globulins and albumins proteins from quinoa seeds retain more water than wheat protein, which indicates that the gluten network from composite dough was diluted and decreased the alfa amylase activity that affects the proofing process. The results were in line with those obtained previously by Park, Maeda, and Morita [51], Wang et al. [19], and Kurek and Sokolova [17]. Only small PS decreased bread volume significantly (p < 0.01), while no significant differences were observed between the large and medium particles on the volume of bread. This may be related to the higher water absorption capacity of small fractions [17,52]. The porosity and elasticity of WF-QF composite flour were affected significantly ($p < 0.01$) by the QF addition level, type of PS, and interaction between them. Porosity and elasticity decreased with the amount QF added, while PS determined an irregular trend.

The color of composite bread crust was significantly ($p < 0.01$) browner compared with the control (Table 5). The addition level of quinoa flour decreased the lightness ($L^*$) of bread crust which could be due to the color intensity of the raw PS flour and to the dark-colored Maillard reaction products on the crust surface. QF had a higher activity of α–amylase (low falling number) [39] than wheat flour and this could explain the darkness (low $L^*$ value) of bread.

**Table 5.** Crust and crumb color parameters of bread samples as affected by quinoa flours fractions.

| Sample | Crust Color | | | Crumb Color | | |
|---|---|---|---|---|---|---|
| | $L^*$ | $a^*$ | $b^*$ | $L^*$ | $a^*$ | $b^*$ |
| Control | 67.36 ± 0.19 [d] | 0.78 ± 0.22 [a] | 32.27 ± 0.28 [cy] | 72.30 ± 0.27 [e] | −4.48 ± 0.03 [a] | 19.02 ± 0.23 [a] |
| QL_5 | 64.99 ± 0.74 [dx] | 4.09 ± 0.30 [by] | 29.65 ± 0.17 [az] | 75.21 ± 0.19 [dy] | −4.24 ± 0.09 [bx] | 19.75 ± 0.13 [bx] |
| QL_10 | 61.79 ± 0.07 [cx] | 5.29 ± 0.10 [cy] | 32.74 ± 0.20 [bz] | 64.64 ± 1.07 [cy] | −3.95 ± 0.24 [cx] | 20.09 ± 0.59 [cx] |
| QL_15 | 60.71 ± 0.40 [bx] | 6.64 ± 0.24 [dy] | 34.00 ± 0.78 [cz] | 64.00 ± 0.50 [by] | −3.86 ± 0.03 [dx] | 21.30 ± 0.56 [dx] |
| QL_20 | 59.88 ± 0.97 [ax] | 6.79 ± 0.53 [dy] | 34.28 ± 0.44 [dz] | 63.37 ± 0.47 [ay] | −3.67 ± 0.04 [ex] | 21.26 ± 0.10 [ex] |
| QM_5 | 63.36 ± 0.56 [by] | 3.59 ± 0.25 [bx] | 24.56 ± 0.22 [ax] | 69.88 ± 0.73 [dy] | −4.17 ± 0.14 [by] | 19.54 ± 0.60 [by] |
| QM_10 | 65.48 ± 0.43 [cy] | 4.74 ± 0.38 [cx] | 30.66 ± 0.59 [bx] | 66.96 ± 0.85 [cy] | −3.78 ± 0.02 [cy] | 21.54 ± 0.22 [cy] |
| QM_15 | 63.62 ± 0.26 [by] | 4.90 ± 0.18 [dx] | 31.32 ± 0.87 [cx] | 65.64 ± 0.38 [by] | −3.33 ± 0.02 [dy] | 21.89 ± 0.07 [dy] |
| QM_20 | 62.23 ± 0.51 [ay] | 5.11 ± 0.32 [dx] | 32.76 ± 0.69 [dx] | 63.40 ± 0.67 [ay] | −3.17 ± 0.09 [ey] | 23.50 ± 0.22 [ey] |
| QS_5 | 64.25 ± 0.31 [dx] | 3.54 ± 0.09 [bx] | 29.03 ± 1.14 [ay] | 65.45 ± 1.27 [dx] | −3.74 ± 0.08 [bz] | 20.42 ± 0.49 [by] |
| QS_10 | 62.01 ± 0.61 [cx] | 4.42 ± 0.20 [cx] | 30.51 ± 0.36 [by] | 65.45 ± 0.33 [cx] | −3.66 ± 0.04 [cz] | 22.12 ± 0.70 [cy] |
| QS_15 | 60.01 ± 0.74 [bx] | 4.78 ± 0.32 [dx] | 31.71 ± 0.43 [cy] | 65.35 ± 0.51 [bx] | −3.03 ±0.09 [dz] | 22.78 ± 0.49 [dy] |
| QS_20 | 57.79 ± 0.88 [ax] | 5.11 ± 0.14 [dx] | 34.57 ± 0.41 [dy] | 64.37 ± 1.69 [ax] | −2.23 ± 0.10 [ez] | 22.19 ± 1.25 [ey] |
| Two-way ANOVA $p$ value | | | | | | |
| F1 | $p < 0.01$ | $p < 0.01$ | $p < 0.01$ | $p < 0.01$ | $p < 0.01$ | $p < 0.01$ |
| F2 | $p < 0.01$ | $p < 0.01$ | $p < 0.01$ | $p < 0.01$ | $p < 0.01$ | $p < 0.01$ |
| F1 × F2 | $p < 0.01$ | $p < 0.01$ | $p < 0.01$ | $p < 0.01$ | $p < 0.01$ | $p < 0.01$ |

F1: level of QF addition; F2: type of particle size; means in the same column with different superscript letters indicate significant difference ($p < 0.01$): [a–e] for QF addition level (0–20%); and [x–z] for QF PS (L, M, and S). $L^*$, $a^*$, $b^*$: CIELAB color parameters.

Redness ($a^*$ value) of the control was lower than bread containing QF, the crust $a^*$ values showing significant differences ($p < 0.01$) between samples, increasing when QF addition increased. Yellowness ($b^*$ value) increased gradually by increasing the substitution levels of QF. Generally, the results showed that the QF bread samples were darker and redder than the control bread sample which was in accordance with El-Sohaimy et al. [18] and Bilgiçli and İbanoğlu [53]. Carotenoids, chlorophyll, and lignin from quinoa seeds influence the color of flour, crumbs, and crust of the products [54].

Bread crumb brightness was significantly ($p < 0.01$) influenced by the QF addition level and PS. The addition level of quinoa flour in composite flour significantly decreased the lightness ($L^*$) of bread crumb, while the type of PS decreased the crumb lightness as follows: M > S > L. The ($a^*$) redness and ($b^*$) yellowness significantly ($p < 0.01$) raised when the addition level of quinoa flour increased and PS decreased. These results are in line with previous work [18] and can be explained by the higher content of protein in quinoa flour than wheat.

### 3.6. Textural Parameters of Bread

Quinoa flour at different PS and addition levels had significant effects on the bread samples' texture profiles (Figure 4). Increasing quinoa substitution of wheat flour and interaction of PS and QF significantly increased the firmness of the bread crumb. The bread sample QL_5 presented a lower firmness value than the control bread, which could be due to albumin protein from quinoa seeds, because it can act like gluten in the dough.

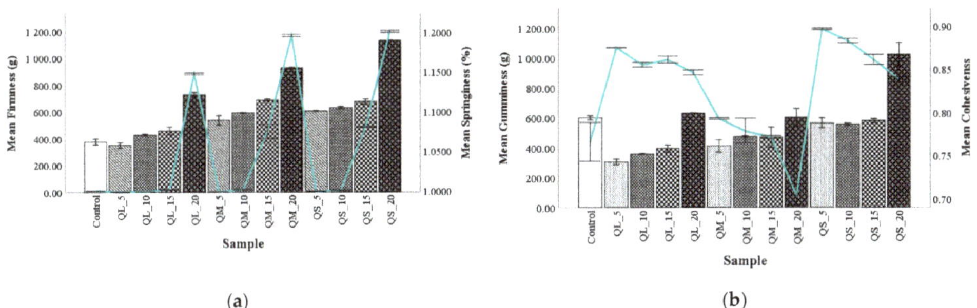

**Figure 4.** The effect of quinoa flours fractions on bread texture: hardness and springiness (**a**), gumminess and cohesiveness (**b**).

It can be observed that all PS at 20% addition significantly increased the firmness of bread, which can be explained by the reduction in the percentage of gluten (responsible for the softness of bread), being also correlated with the high protein content that is found in composite flour which led hard and crunchy bread [18]. The incorporation of 20% QF could cause a negative impact on the acceptability of the bread. This could be related to the significant reduction of air-retention ability and specific volume of the breads [19]. These findings are in line with the results found by Wang et al. [19], El-Sohaimy, Shehata, Mehany, and Zeitoun [22], and by Wolter et al. [55]. Bread springiness followed the same trend as firmness, which raised when more that 10% QF was incorporated. Gumminess increased when QF addition level increased and with the decrease of PS, which may be related to the high content of protein and dietary fiber in quinoa blends, similar results being observed by El-Sohaimy et al. [18]. Cohesiveness in wheat flour and 5% quinoa flour for all PS were slightly higher than in other composite flour, but in general, it decreased with the increase of QF addition and it was higher when PS decreased. These variations could be explained due to the presence of prolamins contained gliadin from wheat flour.

### 3.7. Relations between the Characteristics

The Pearson correlation coefficients (0.56 > r < 0.99) were determined between composite flour chemical constituents, dough rheological and textural parameters, and bread characteristics. Between flour moisture and dough tenacity L (r = 0.75), dough extensibility G (r = 0.76), dough strength W (r = 0.88), dough adhesiveness (r = 0.82), tan δ (r = 0.89), and bread volume (r = 0.97) were found significant positive correlations, while the moisture content was negatively correlated with dough hardness (r = −0.83) and bread firmness (r = −0.82). Probably, in this case, a certain quantity of water enhances the viscoelastic behavior of dough, this amount of water being necessary for protein swelling, the best dough consistency being obtained when enough water is used to swell composite flour components. Similar positive and negative correlations were found between carbohydrates and the parameters listed above.

High positive correlations were found between lipids from flour and dough hardness (r = 0.76) and with bread firmness (r = 0.75). The lipids content from composite flour was negatively correlated with dough extensibility L (r = −0.76), and dough strength W (r = −0.86), dough adhesiveness (r = −0.86), tan δ (r = − 0.84), and with bread elasticity (r = −0.72), volume (r = −0.96), and porosity (r = −0.63). Lipids had a significant influence on bread texture and quality due to their capacity to associate with proteins as they present hydrophilic and hydrophobic groups, and with starch, resulting in starch-lipid complexes [56]. The same trend of positive and negative correlations within these parameters were found with ash content of composite flour. Additionally, a high positive correlation was found between $T_{max}$ and dough extensibility L (r = 0.83), and W (r = 0.76), while negative relationships were found between dough biaxial measurements and G′ and G″. High positive correlations were found between bread volume and L (r = 0.75), G (r = 0.76), W (r = 0.87), dough adhesiveness (r = 0.82), and tan δ (r = 0.88), but a negative relationship between dough rheological parameters and bread physical parameters: dough hardness with bread elasticity (r = −0.74), bread volume, and bread firmness (r = −0.74) were observed. All the correlations listed above were significant at $p < 0.05$. Similar correlation for flour chemical constituents, dough, and bread parameters were found by other authors [57,58].

The principal component analysis (PCA) was used to put in evidence the effect of QF addition and PS on wheat-quinoa composite flour, dough, and bread variables (Figure 5). The two principal components explained 73.03% of the total variance (PC1 = 52.85% and PC2 = 20.18%). The PC1 was associated with composite flour moisture, lipids, ash, carbohydrates, dough alveographic parameters (L, G, W, and P/L), dough textural parameters (hardness, adhesiveness), elastic modulus (G′), tan δ, and bread physical properties (elasticity, volume), while PC2 was associated with dough tenacity (P), viscous modulus (G″) and bread gumminess. It can be observed a high opposition between protein and carbohydrates, P and L alveograph parameters, bread elasticity, and dough hardness.

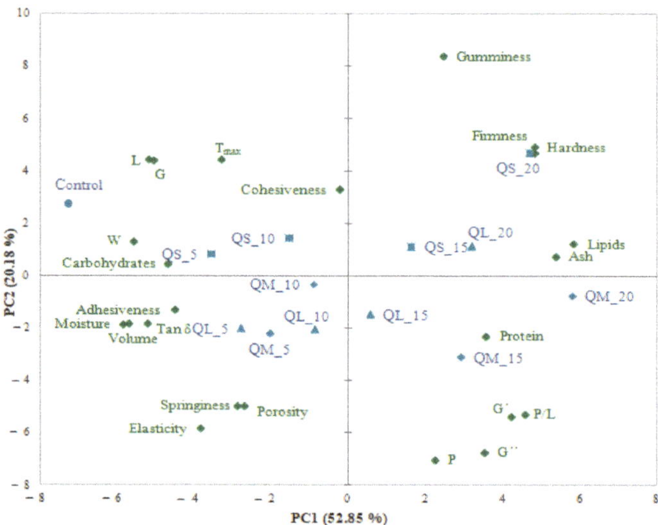


**Figure 5.** Principal component analysis bi-plot: distribution of the proximate composition, dough rheological and textural parameters, bread physical and textural parameters, and samples.

Regarding bread samples, a good relationship can be observed between the control sample and bread with a 5–10% QF addition level. Samples with medium PS (15–20%) were associated with protein, in opposition to the samples with 20% quinoa large and small fractions which were associated with dough and bread hardness.

## 4. Conclusions

The addition of quinoa flour induced significant changes in dough rheological and textural parameters, bread color, texture, and physical properties, depending on the addition level and particle size used. Quinoa fractionation determined different structural and molecular characteristics, depending on the particle size. The composite flours showed higher chemical components contents in terms of proteins (≥12.47%), lipids (≥1.65%), and ash (≥0.70%), while the carbohydrates content, which varied from 70.20% to 71.25%, was lower compared with the control (71.36%). Higher dough tenacity, hardness and dynamic moduli, and lower extensibility, adhesiveness and dough strength were obtained as the addition level was higher. Bread with raised firmness, springiness and gumminess, was obtained as the quinoa flour level was higher. Bread volume, porosity, elasticity, and luminosity decreased from 378.70 cm$^3$ to 260.00 cm$^3$, from 72.38% to 57.27%, and from 97.92%

to 87.72%, respectively, when QF was added. In order to achieve the highest technological and quality characteristics of bread enriched with quinoa fractions, an optimization of the processing parameters could be performed, taking into account the producers' conditions and desires. The results presented in this study suggested that the addition levels of 5–15% quinoa fractions to wheat flour could provide acceptable quality characteristics such as rheological behavior which may predict dough handling during processing and final product color, texture, and physical properties. Thus, these results could be of interest for processors in order to develop novel bread formulation with superior characteristics and increased nutritional value.

**Supplementary Materials:** The following are available online at https://www.mdpi.com/article/10.3390/plants10102150/s1, Figure S1: Variations of elastic (G′) and viscous modulus (G″) with frequency for wheat-quinoa fractions with large (L), medium (M), and small (S) particle sizes dough with: 5% (a), 10% (b), 15% (c), and 20% (d) addition level; Figure S2: Variations of elastic (G′) and viscous modulus (G″) with temperature for wheat-quinoa fractions with large (L), medium (M), and small (S) particle sizes dough with: 5% (a), 10% (b), 15% (c), and 20% (d) addition level.

**Author Contributions:** The authors contributed equally to this research and approved the final version of the manuscript. All authors have read and agreed to the published version of the manuscript.

**Funding:** This work did not receive external funding.

**Institutional Review Board Statement:** Not applicable.

**Informed Consent Statement:** Not applicable.

**Data Availability Statement:** Not applicable.

**Acknowledgments:** This work was supported by Romania National Council for Higher Education Funding, CNFIS, project number CNFIS-FDI-2021-0357.

**Conflicts of Interest:** The authors declare no conflict of interest.

# References

1. Vignola, M.B.; Moiraghi, M.; Salvucci, E.; Baroni, V.; Pérez, G.T. Whole meal and white flour from Argentine wheat genotypes: Mineral and arabinoxylan differences. *J. Cereal Sci.* **2016**, *71*, 217–223. [CrossRef]
2. Mironeasa, S.; Iuga, M.; Zaharia, D.; Mironeasa, C. Optimization of grape peels particle size and flour substitution in white wheat flour dough. *Sci. Study Res. Chem. Chem. Eng. Biotechnol. Food Ind.* **2019**, *20*, 29–42.
3. Saleem Khan, M.; Ali, E.; Ali, S.; Khan, W.M.; Sajjad, M.A.; Hussain, F.; Khan, M.S.; Ali, S.; Hussain, F. Assessment of essential amino acids in wheat proteins: A case study. *J. Biodivers. Environ. Sci.* **2014**, *4*, 185–189.
4. Küster, I.; Vila, N. Health/Nutrition food claims and low-fat food purchase: Projected personality influence in young consumers. *J. Funct. Foods* **2017**, *38*, 66–76. [CrossRef]
5. Encina-Zelada, C.; Cadavez, V.; Pereda, J.; Gómez-Pando, L.; Salvá-Ruíz, B.; Teixeira, J.A.; Ibañez, M.; Liland, K.H.; Gonzales-Barron, U. Estimation of composition of quinoa (*Chenopodium quinoa* Willd.) grains by Near-Infrared Transmission spectroscopy. *LWT* **2017**, *79*, 126–134. [CrossRef]
6. Navruz-Varli, S.; Sanlier, N. Nutritional and health benefits of quinoa (*Chenopodium quinoa* Willd.). *J. Cereal Sci.* **2016**, *69*, 371–376. [CrossRef]
7. Nowak, V.; Du, J.; Charrondière, U.R. Assessment of the nutritional composition of quinoa (*Chenopodium quinoa* Willd.). *Food Chem.* **2016**, *193*, 47–54. [CrossRef]
8. Valcárcel-Yamani, B.; Lannes, S.C.d.S. Applications of quinoa (*Chenopodium quinoa* Willd.) and amaranth (*Amaranthus* spp.) and their influence in the nutritional value of cereal based foods. *Food Public Health* **2012**, *2*, 265–275.
9. Pereira, E.; Encina-Zelada, C.; Barros, L.; Gonzales-Barron, U.; Cadavez, V.; Isabel, C.F.R.; Ferreira, I. Chemical and nutritional characterization of *Chenopodium quinoa* Willd (quinoa) grains: A good alternative to nutritious food. *Food Chem.* **2019**, *280*, 110–114. [CrossRef]
10. Li, L.; Lietz, G.; Seal, C.J. Phenolic, apparent antioxidant and nutritional composition of quinoa (*Chenopodium quinoa* Willd.) seeds. *Int. J. Food Sci. Technol.* **2021**, *56*, 3245–3254. [CrossRef]
11. Prego, I.; Maldonado, S.; Otegui, M. Seed structure and localization of reserves in *Chenopodium quinoa*. *Ann. Bot.* **1998**, *82*, 481–488. [CrossRef]
12. Ando, H.; Chen, Y.C.; Tang, H.; Shimizu, M.; Watanabe, K.; Mitsunaga, T. Food Components in Fractions of Quinoa Seed. *Food Sci. Technol. Res.* **2002**, *8*, 80–84. [CrossRef]

13. Tsatsaragkou, K.; Yiannopoulos, S.; Kontogiorgi, A.; Poulli, E.; Krokida, M.; Mandala, I. Mathematical approach of structural and textural properties of gluten free bread enriched with carob flour. *J. Cereal Sci.* **2012**, *56*, 603–609. [CrossRef]
14. Cappelli, A.; Cini, E.; Guerrini, L.; Masella, P.; Angeloni, G.; Parenti, A. Predictive models of the rheological properties and optimal water content in doughs: An application to ancient grain flours with different degrees of refining. *J. Cereal Sci.* **2018**, *83*, 229–235. [CrossRef]
15. Solaesa, Á.G.; Villanueva, M.; Vela, A.J.; Ronda, F. Protein and lipid enrichment of quinoa (cv.Titicaca) by dry fractionation. Techno-functional, thermal and rheological properties of milling fractions. *Food Hydrocoll.* **2020**, *105*, 105770. [CrossRef]
16. Mironeasa, S.; Zaharia, D.; Codină, G.G.; Ropciuc, S.; Iuga, M. Effects of Grape Peels Addition on Mixing, Pasting and Fermentation Characteristics of Dough from 480 Wheat Flour Type. *Bull. UASVM Food Sci. Technol.* **2018**, *75*, 27–35. [CrossRef]
17. Kurek, M.A.; Sokolova, N. Optimization of bread quality with quinoa flour of different particle size and degree of wheat flour replacement. *Food Sci. Technol.* **2020**, *40*, 307–314. [CrossRef]
18. El-Sohaimy, A.S.; Shehata, G.M.; Djapparovec, T.A.; Mehany, T.; Zeitoun, A.M.; Zeitoun, M.A. Development and characterization of functional pan bread supplemented with quinoa flour. *J. Food Process. Preserv.* **2021**, *45*. [CrossRef]
19. Wang, X.; Lao, X.; Bao, Y.; Guan, X.; Li, C. Effect of whole quinoa flour substitution on the texture and in vitro starch digestibility of wheat bread. *Food Hydrocoll.* **2021**, *119*, 106840. [CrossRef]
20. Calderelli, V.A.S.; de Benassi, M.T.; Visentainer, J.V.; Matioli, G. Quinoa and flaxseed: Potential ingredients in the production of bread with functional quality. *Brazilian Arch. Biol. Technol.* **2010**, *53*, 981–986. [CrossRef]
21. Stikic, R.; Glamoclija, D.; Demin, M.; Vucelic-Radovic, B.; Jovanovic, Z.; Milojkovic-Opsenica, D.; Jacobsen, S.-E.; Milovanovic, M. Agronomical and nutritional evaluation of quinoa seeds (*Chenopodium quinoa* Willd.) as an ingredient in bread formulations. *J. Cereal Sci.* **2012**, *55*, 132–138. [CrossRef]
22. El-Sohaimy, S.A.; Shehata, M.G.; Mehany, T.; Zeitoun, M.A. Nutritional, Physicochemical, and Sensorial Evaluation of Flat Bread Supplemented with Quinoa Flour. *Int. J. Food Sci.* **2019**, *2019*. [CrossRef]
23. Xu, X.; Luo, Z.; Yang, Q.; Xiao, Z.; Lu, X. Effect of quinoa flour on baking performance, antioxidant properties and digestibility of wheat bread. *Food Chem.* **2019**, *294*, 87–95. [CrossRef] [PubMed]
24. Coțovanu, I.; Mironeasa, S. Impact of different amaranth particle size addition level on wheat flour dough rheology and bread features. *Foods* **2021**, *10*, 1539. [CrossRef] [PubMed]
25. Liu, J.; Wang, Z.; Wang, Z.; Hao, Y.; Wang, Y.; Yang, Z.; Li, W.; Wang, J. Physicochemical and functional properties of soluble dietary fiber from different colored quinoa varieties (*Chenopodium quinoa* Willd). *J. Cereal Sci.* **2020**, *95*, 103045. [CrossRef]
26. Shotts, M.L.; Plans Pujolras, M.; Rossell, C.; Rodriguez-Saona, L. Authentication of indigenous flours (Quinoa, Amaranth and kañiwa) from the Andean region using a portable ATR-Infrared device in combination with pattern recognition analysis. *J. Cereal Sci.* **2018**, *82*, 65–72. [CrossRef]
27. Czekus, B.; Pećinar, I.; Petrović, I.; Paunović, N.; Savić, S.; Jovanović, Z.; Stikić, R. Raman and Fourier transform infrared spectroscopy application to the Puno and Titicaca cvs. of quinoa seed microstructure and perisperm characterization. *J. Cereal Sci.* **2019**, *87*, 25–30. [CrossRef]
28. Mironeasa, S.; Codină, G.G.; Mironeasa, C. Effect of Composite Flour Made from Tomato Seed and Wheat of 650 Type of a Strong Quality for Bread Making on Bread Quality and Alveograph Rheological Properties. *Int. J. Food Eng.* **2018**, *4*, 22–26. [CrossRef]
29. Mironeasa, S.; Iuga, M.; Zaharia, D.; Mironeasa, C. Rheological Analysis of Wheat Flour Dough as Influenced by Grape Peels of Different Particle Sizes and Addition Levels. *Food Bioprocess Technol.* **2019**, *12*, 228–245. [CrossRef]
30. Iuga, M.; Boestean, O.; Ghendov-Mosanu, A.; Mironeasa, S. Impact of Dairy Ingredients on Wheat Flour Dough Rheology and Bread Properties. *Foods* **2020**, *9*, 828–854. [CrossRef]
31. Mironeasa, S. Mironeasa Costel Dough bread from refined wheat flour partially replaced by grape peels: Optimizing the rheological properties. *J. Food Process Eng.* **2019**, *42*, 1–14. [CrossRef]
32. Bordei, D.; Bahrim, G.; Pâslaru, V.; Gasparotti, C.; Elisei, A.; Banu, I.; Ionescu, L.; Codină, G. Quality Control in the Bakery Industry-Analysis Methods. *Galați Acad.* **2007**, *1*, 203–212.
33. Romano, A.; Masi, P.; Nicolai, M.A.; Falciano, A.; Ferranti, P. QUINOA (*Chenopodium quinoa* willd.) flour as novel and safe ingredient in bread formulation. *Chem. Eng. Trans.* **2019**, *75*, 301–306. [CrossRef]
34. Alvarez-Jubete, L.; Auty, M.; Arendt, E.K.; Gallagher, E. Baking properties and microstructure of pseudocereal flours in gluten-free bread formulations. *Eur. Food Res. Technol.* **2010**, *230*, 437–445. [CrossRef]
35. Marti, A.; Bock, J.E.; Pagani, M.A.; Ismail, B.; Seetharaman, K. Structural characterization of proteins in wheat flour doughs enriched with intermediate wheatgrass (*Thinopyrum intermedium*) flour. *Food Chem.* **2016**, *194*, 994–1002. [CrossRef]
36. Cozzolino Daniel The role of visible and infrared spectroscopy combined with chemometrics to measure phenolic compounds in grape and wine samples. *Molecules* **2015**, *20*, 726–737. [CrossRef]
37. García-Salcedo, Á.J.; Torres-Vargas, O.L.; Ariza-Calderón, H. Physical-chemical characterization of quinoa (*Chenopodium quinoa* Willd.), amaranth (*Amaranthus caudatus* L.), and chia (*Salvia hispanica* L.) flours and seeds. *Agroindustry Food Sci.* **2017**, *67*, 215–222.
38. Prasopsunwattana, N.; Omary, M.B.; Arndt, E.A.; Cooke, P.H.; Flores, R.A.; Yokoyama, W.; Toma, A.; Chongcham, S.; Lee, S.P. Particle size effects on the quality of flour tortillas enriched with whole grain waxy barley. *Cereal Chem.* **2009**, *86*, 439–451. [CrossRef]
39. Coțovanu, I.; Stoenescu, G.; Mironeasa, S. Amaranth Influence on Wheat Flour Dough Rheology: Optimal Particle Size and Amount of Flour Replacement. *J. Microbiol. Biotechnol. Food Sci.* **2020**, *10*, 366–373. [CrossRef]

40. Chauhan, G.S.; Zillman, R.R.; Eskin, N.A.M. Dough mixing and breadmaking properties of quinoa-wheat flour blends. *Int. J. Food Sci. Technol.* **1992**, *27*, 701–705. [CrossRef]
41. Dhen, N.; Román, L.; Ben Rejeb, I.; Martínez, M.M.; Garogouri, M.; Gómez, M. Particle size distribution of soy flour affecting the quality of enriched gluten-free cakes. *LWT* **2016**, *66*, 179–185. [CrossRef]
42. Ahmed, J.; Thomas, L.; Arfat, Y.A. Functional, rheological, microstructural and antioxidant properties of quinoa flour in dispersions as influenced by particle size. *Food Res. Int.* **2019**, *116*, 302–311. [CrossRef]
43. Demir, M.K. Use of quinoa flour in the production of gluten-free tarhana. *Food Sci. Technol. Res.* **2014**, *20*, 1087–1092. [CrossRef]
44. Gallagher, E.; Gormley, T.R.; Arendt, E.K. Crust and crumb characteristics of gluten free breads. *J. Food Eng.* **2003**, *56*, 153–161. [CrossRef]
45. Sluimer, P.; Sluimer, P. *Principles of Breadmaking: Functionality of Raw Materials and Process Steps*; American Association of Cereal Chemists (AACC): St. Paul, MN, USA, 2005.
46. Coțovanu, I.; Mironeasa, S. Buckwheat seeds: Impact of milling fractions and addition level on wheat bread dough rheology. *Appl. Sci.* **2021**, *11*, 1731. [CrossRef]
47. Hoseney, R.C. *Principles of Cereal Science and Technology*; American Association of Cereal Chemists (AACC): St. Paul, MN, USA, 1994; ISBN 0913250791.
48. Ratnawati, L.; Desnilasari, D.; Surahman, D.N.; Kumalasari, R. Evaluation of Physicochemical, Functional and Pasting Properties of Soybean, Mung Bean and Red Kidney Bean Flour as Ingredient in Biscuit. *IOP Conf. Ser. Earth Environ. Sci.* **2019**, *251*. [CrossRef]
49. Tamba-Berehoiu, R.M.; Turtoi, M.O.; Popa, C.N. Assessment of quinoa flours effect on wheat flour dough rheology and bread quality. *Ann. Univ. Dunarea Jos Galati Fascicle VI Food Technol.* **2019**, *43*, 173–188. [CrossRef]
50. Föste, M.; Nordlohne, S.D.; Elgeti, D.; Linden, M.H.; Heinz, V.; Jekle, M.; Becker, T. Impact of quinoa bran on gluten-free dough and bread characteristics. *Eur. Food Res. Technol.* **2014**, *239*, 767–775. [CrossRef]
51. Park, S.H.; Maeda, T.; Morita, N. Effect of Whole Quinoa Flours and Lipase on the Chemical, Rheological and Breadmaking Characteristics of Wheat Flour. *J. Appl. Glycosci.* **2005**, *52*, 337–343. [CrossRef]
52. Coțovanu, I.; Batariuc, A.; Mironeasa, S. Characterization of quinoa seeds milling fractions and their effect on the rheological properties of wheat flour dough. *Appl. Sci.* **2020**, *10*, 7225. [CrossRef]
53. Bilgiçli, N.; İbanoğlu, Ş. Effect of pseudo cereal flours on some physical, chemical and sensory properties of bread. *J. Food Sci. Technol.* **2015**, *52*, 7525–7529. [CrossRef]
54. Ruffino, A.M.C.; Rosa, M.; Hilal, M.; González, J.A.; Prado, F.E. The role of cotyledon metabolism in the establishment of quinoa (*Chenopodium quinoa*) seedlings growing under salinity. *Plant Soil* **2010**, *326*, 213–224. [CrossRef]
55. Wolter, A.; Hager, A.S.; Zannini, E.; Arendt, E.K. In vitro starch digestibility and predicted glycaemic indexes of buckwheat, oat, quinoa, sorghum, teff and commercial gluten-free bread. *J. Cereal Sci.* **2013**, *58*, 431–436. [CrossRef]
56. Goesaert, H.; Brijs, K.; Veraverbeke, W.S.; Courtin, C.M.; Gebruers, K.; Delcour, J.A. Wheat flour constituents: How they impact bread quality, and how to impact their functionality. *Trends Food Sci. Technol.* **2005**, *16*, 12–30. [CrossRef]
57. Miñarro, B.; Albanell, E.; Aguilar, N.; Guamis, B.; Capellas, M. Effect of legume flours on baking characteristics of gluten-free bread. *J. Cereal Sci.* **2012**, *56*, 476–481. [CrossRef]
58. Elgeti, D.; Nordlohne, S.D.; Föste, M.; Besl, M.; Linden, M.H.; Heinz, V.; Jekle, M.; Becker, T. Volume and texture improvement of gluten-free bread using quinoa white flour. *J. Cereal Sci.* **2014**, *59*, 41–47. [CrossRef]

*Article*

# Comparative Study of Volatile Compounds and Sensory Characteristics of Dalmatian Monovarietal Virgin Olive Oils

Mirella Žanetić [1,*], Maja Jukić Špika [1,2], Mia Mirjana Ožić [3] and Karolina Brkić Bubola [4]

[1] Institute of Adriatic Crops and Karst Reclamation, Put Duilova 11, HR-21000 Split, Croatia; Maja.Jukic.Spika@krs.hr

[2] Centre of Excellence for Biodiversity and Molecular Plant Breeding (CoE CroP-BioDiv), Svetošimunska Cesta 25, HR-10000 Zagreb, Croatia

[3] Education and Teacher Training Agency, Tolstojeva 32, HR-21000 Split, Croatia; mmozic@azoo.hr

[4] Institute of Agriculture and Tourism, K. Huguesa 8, HR-52440 Poreč, Croatia; karolina@iptpo.hr

* Correspondence: Mirella.Zanetic@krs.hr; Tel.: +385-21-434-418

**Abstract:** Volatile compounds are chemical species responsible for the distinctive aroma of virgin olive oil. Monovarietal olive oils have a peculiar composition of volatiles, some of which are varietal descriptors. In this paper, the total phenolic content (TPC), fatty acid composition, volatile compounds, and sensory profile of monovarietal olive oils from four Dalmatian most common olive cultivars—Oblica, Lastovka, Levantinka, and Krvavica—were studied. The volatile composition of olive oils was analyzed using headspace solid-phase microextraction with gas chromatography/mass spectrometry. The highest mean TPC value was measured in Oblica and Krvavica oils (around 438 mg/kg). The difference among cultivars for fatty acids composition was detected for C16:1, C17:0, C18:1, C18:2, and the ratio C18:1/C18:2. Krvavica oils showed clear differences in fatty acid composition compared to oils from other cultivars. The most prevalent volatile compound in all oils was C6 aldehyde E-2-hexenal, with the highest value detected in Levantinka oils (75.89%), followed by Lastovka (55.27%) and Oblica (54.86%). Oblica oils had the highest value of Z-3-hexen-1-ol, which influenced its characteristic banana fruitiness, detected only in this oil. Lastovka oils had the highest amount of several volatiles (heptanal, Z-2-heptenal, hexanal, hexyl acetate), with a unique woody sensation and the highest astringency among all studied cultivars. Levantinka oils had the highest level of almond fruitiness, while Krvavica oils had the highest level of grass fruitiness.

**Keywords:** virgin olive oil; *Olea europea* L.; phenols; sensory profile; fatty acid composition; volatile compounds

**Citation:** Žanetić, M.; Jukić Špika, M.; Ožić, M.M.; Brkić Bubola, K. Comparative Study of Volatile Compounds and Sensory Characteristics of Dalmatian Monovarietal Virgin Olive Oils. *Plants* 2021, *10*, 1995. https://doi.org/10.3390/plants10101995

Academic Editor: Ivo Vaz de Oliveira

Received: 25 August 2021
Accepted: 21 September 2021
Published: 23 September 2021

**Publisher's Note:** MDPI stays neutral with regard to jurisdictional claims in published maps and institutional affiliations.

## 1. Introduction

Food selection generally requires fulfilling fundamental nutritional needs: meeting basic nutrient content, improving physical and mental health, and reducing the risk of common diseases. Virgin olive oil (VOO) claims all of these features [1]. In addition, the food must also be delicious in order to be appealing to the consumer. The pleasant taste and aroma of VOO are the main attributes that guide consumers to choose this product among other vegetable oils. As a unique fat ingredient of the Mediterranean diet, VOO is well known as a vegetable oil produced directly from fresh olive fruit, processed mechanically under a strictly controlled temperature regime; it is consumed in an unrefined state [2]. For this reason, VOO keeps its original composition very similar to the one present in olive fruit while still being kept inside vacuoles. Its nutraceutical effect is due to its particular chemical composition since it consists of mainly a monounsaturated fatty acid (MUFA), namely, oleic acid (which controls cholesterol levels), and essential fatty acids, namely, linoleic and α-linolenic acids (which lower the risk of coronary diseases and the incidence of different types of cancers) [3]. Oxidation of VOO takes place in unsaturated fatty acids, so the main substrates in the olive oil's glyceride part are oleic acid (about 55–83%) [4] and double unsaturated linoleic acid (3.5–21%), while triple unsaturated linolenic acid is present with less than 1% [4,5]. International standards [4,5] establish the limits for the purity criteria of olive oils and olive pomace oils, and the proportions of single fatty acids are one of the most important criterion parameters. Both mentioned regulations have the limit values for VOO categorization that are mutually harmonized. Furthermore, VOO is rich in valuable amounts of important bioactive compounds, mainly tocopherols, phenols, sterols, hydrocarbons, and carotenoids [3,6]. Many of these compounds are natural antioxidants that have a protective role against oxidative deprivation during oil storage as well as guard the human body against different diseases [6]. Moreover, phenolic compounds are responsible for the bitterness and pungency sensation, two positive and desirable sensory attributes in VOO. The chemical composition of monovarietal VOO intensely depends on agronomic factors and the olive cultivar (genetic origin) and also defines its volatile profile [6–12]. For the unique VOO aroma, the volatile compounds formed during the crushing of olives, the malaxation of olive paste, and after the extracting process of oil are principally responsible. This group of compounds involves mainly C5 and C6 units formed from polyunsaturated fatty acids during the enzymatic process of the lipoxygenase pathway [13,14] and other consequent bioprocesses through a technological oil extraction process. There is a positive correlation between the concentration of phenols and volatile compounds produced by LOX pathways and the intensity of single sensory properties [15]. This mechanism shows a strong link between essential fatty acids and volatile compounds that are formed via the enzymatic activity of lipoxygenase, hydroperoxydelyase, and other enzymes involved in this reaction. The synthesis of VOO aroma compounds is a subject of interest for many research groups [13–16].

The volatile fraction of VOO consists of five main groups of compounds: aldehydes, alcohols, esters, terpenes, and organic acids. The most common VOO volatile compounds are: C6 aldehydes (hexanal, Z-3-hexenal, E-2-hexenal), representing 60–80%; C6 alcohols (hexanol, Z-3-hexenol, E-2-hexenol); and C6 esters (hexyl acetate, Z-3-hexenil acetate) [10]. The volatile fraction has a particular composition regarding olive cultivars and can be used for the characterization of monovarietal VOOs [17–24]. Olive (*Olea europaea* L.) cultivars present a high variability for VOO volatile compounds and, similarly, of the aroma quality. This natural variation of aroma profiling is a result of the high olive genetic diversity [25–27], but also it depends on other factors such is fruit ripening degree [24,28,29], irrigation regime in the orchard [30], fruit processing conditions during oil extraction [31,32], and correct olive oil storage practice [33–36]. Different pre- and post-harvest factors influence the olive oil composition and quality of olive fruit [37]. The olive cultivar influences the physico-chemical and sensorial properties of the oil. An important impact of olive cultivar is on volatile compounds [10] and fatty acid composition [38]. The olive oil flavor depends on its area of geographical origin because it is influenced by environmental conditions [10]. The growing area has impact on cis-3-hexenal, cis-3-hexenol, hexanal, hexanol, trans-2-hexenal, trans-3-hexenol and trans-2-hexenol [38]. The geographical cultivation area was also found to influence fatty acid composition [39]. At colder locations of higher altitude, both studied cultivars (Oblica and Leccino) had higher amounts of stearic, linoleic, and linolenic fatty acids as well as a higher proportion of phenolic compounds but lower amounts of oleic fatty acids. At warmer locations of lower altitude, both cultivars had oils with lower levels of fruitiness, bitterness, and pungency. During olive ripening, oleic and linoleic acids increase and palmitic acid decreases because of different enzymatic activities [40]. Olive fruit maturation degrees play a crucial role in chemical and sensory elements [41]. Olive ripeness affects the volatile compound concentrations [42]. During the early maturation stage, olive fruits show higher volatile compound concentrations compared with the late stage of maturation. Fatty acid concentration is affected by the ripening of the olive fruit, and its concentration differs depending upon the variety [37]. Water deficits in olive orchards usually affect the sensory properties of olive oil, mainly in the high intensity of bitterness [41]. Water stress reduces olive production by reducing the endogenous esterases in fruit, while more water availability decreases volatile compounds (trans-2-hexenal, cis-3-hexen-1-ol, and hexanol) in the fruit [37]. Prolonged olive fruit storage between harvest and processing has a negative effect on the produced olive oil because of the possible hydrolyses of triglycerides to free fatty acids by lipase action. For this reason, it is advisable to process olive fruits as soon as possible (within 24 h after harvesting) [36]. The olive technology process has a strong influence on the overall quality of the obtained olive oil. The crushing methods influence the sensory element of olive oil [41]. Prolonged crushing time may degrade the quality of the olive oil. The milling system has an impact on oil quality as well as the sensory properties [37]. Prolonged malaxation generates higher temperatures of olive paste that negatively affect the quality and sensory characteristics of olive oil [37]. The malaxation process generates the olive oil aroma if the concentration of oxygen and phenolic compounds decreases due to the enzymatic oxidation of polyphenol oxidase and peroxidase [31]. Nowadays, modern centrifuge systems are equipped with two- or three-phase extractors. Two-phase systems reduce or eliminate the use of water in the process, with the double advantage of limiting the use of water and reducing or eliminating the production of wastewater. On the contrary, three-phase systems use a certain quantity of water to adjust olive paste density and ease oil extraction; in contrast, there is a significant loss of phenolic compounds with vegetable water [36]. Systems with two and a half phases are the most recent type of decanter that requires the addition of a reduced quantity of water and separates three fractions (oil must, wastewater, and wet pomace). The advantage of this system is a small quantity of wastewater and a greater quantity of phenolic compounds that remains in the oil [36]. Each of the olive processing technological systems comprises a temperature under 27 °C in order to retain olive oil quality and chemical composition as well as sensory characteristics. The proper storage of freshly produced high-quality extra virgin olive oil is indispensable in order to keep its peculiar characteristics for a longer period. The filtration method, light exposure to light, contact with oxygen or higher temperatures, and the trace elements that promote lipid oxidation also shorten the shelf life and sensory properties of olive oil [37]. Volatile compounds from VOO are also influenced by heating during cooking [43]. Comparing volatile profiles of extra virgin olive oil, olive pomace oil, soybean oil, and palm oil in different heating conditions, Giuffre et al. [43] found that heating completely changed the volatile organic compound content of all four studied edible vegetable oils. Extra virgin olive oils showed the highest number of components, with E$-$2-hexenal as the highest in quantity in fresh oil and Z-2-decenal and 2-undecanal as the highest in quantity in heated extra virgin olive oils. By the results of this study, the consumer can decide what temperature and heating time to apply in order to preserve the flavors and to reduce off-flavors that are produced during the heating of extra virgin olive oils and other studied edible oils [43]. The optimization of analytical methods applied for the determination of volatile composition is also a very important step that influences the aroma fingerprint of certain monovarietal VOOs [42,44–49]. The principal sensory characteristics appraised during the organoleptic assessment of VOOs are peculiar flavor features that are generated from volatile and phenolic compounds [50]. In the recent decade, particular research attention has been given to investigate the relationship between aroma compounds and specific VOO attributes perceived by panel groups, including the development of electronic noses [48].

In a recently published review [47], it was emphasized that, today, there are around 700 detected volatile compounds that contribute to the unique 'green and fruity' aroma of VOOs. Many authors have studied the relationship between sensory evaluation and the volatile profile of VOOs [49,51–53], analyzing the composition of aroma compounds responsible for both positive and negative sensory attributes in VOOs. The association of positive sensory properties with the presence of certain volatile compounds by chemometric approach in Italian VOOs confirmed the presence of 16 volatile compounds, 8 of which affect the green sensation, with the remaining 8 affecting sweetness [51]. Positive sensory characteristics were associated with C5 and C6 compounds, and the most abundant ones contributing positively to the aroma profile of VOOs are C5 and C6 aldehydes and alcohols [51].

Analyzing Italian and Spanish monovarietal VOOs, correlations were detected between the major volatile compounds (sum of aldehydes C6) and the orthonasal perception of olive fruitiness and the retronasal odor of almond [52]. Although modern analytical methods are needed to achieve the correct classification of olive oil, it is inevitable to connect the analytical detection based on the identification and quantification of volatile constituents of individual components and the results of panel group sensory evaluation. The latter is due to the complexity of education, training process, and skill [52].

Consumer ability to recognize the positive and negative sensorial attributes of VOOs that are related to their composition has been the research focus of different research groups [47,53–55]. Barbieri et al. [53] studied the ability of average consumers to distinguish olive oils by sensory properties that scientists label as 'healthier'. However, this research has shown their greater susceptibility for 'sweet' aromas versus 'bitter' ones that assume health impact, which indicates an existing space and need for continuous consumer education. Besides positive sensory attributes, the presence of which are obligatory in VOOs, there is always a possibility that some undesirable negative attribute appears due to mistakes in fruit conservation before processing, inadequate oil processing, or/and storage conditions [50].

Croatia has a century-long olive growing tradition, and today, there are more than 40 autochthonous olive cultivars grown mainly in extensive olive orchards [54,56]. Although the production of olive oil is rather modest and represented mainly by small family olive farms, Croatian olive oils are becoming more and more eminent in the world due to their high quality and numerous awards at various international competitions. Olive and olive oil production in Croatia is constantly increasing. According to IOC data from 2012 [57], annual olive oil production in Croatia is about 5500 t, while table olive production is about 1500 t. According to the Central Bureau of Statistics, in 2017, in the Republic of Croatia, 28,947 tons of olives were produced and about 37,463 hL of olive oil. According to the same data, the area under olive groves in 2017 was 18,683 ha, with approximately 5.5 million olive trees [58]. The largest areas under olive trees are located in Split-Dalmatia County. About 50% of production is domestic autochthonous varieties. The predominant olive cultivar is Oblica, the autochthonous olive variety that covers about 75% of all olive orchards [58].

Although the research on Croatian olive oils done so far has been focused mainly on oil characterization based on fatty acid composition, phenols, volatile compounds, sterols, and sensory quality [29,32,59–62], the Dalmatian monovarietal VOOs' volatiles and their sensory profiles have not been sufficiently studied. Dalmatia is a coastal part of Croatia, and it has traditionally been an olive-growing region for more than 2000 years [32]. The knowledge of the particular sensorial characteristics of monovarietal Dalmatian olive oils is very important for a better understanding of their specificity, which contributes to the creation of blends with intended sensorial profiles. In this paper, we study the composition of volatile compounds and the sensory characteristics of monovarietal VOOs obtained from the three most represented autochthonous Dalmatian olive cultivars: Oblica, Lastovka, Levantinka, and the lesser-known cultivar Krvavica; some of them are poorly investigated. Moreover, fatty acid composition and total phenol content of analyzed monovarietal VOOs were observed in order to find existing links between them and aroma volatile compounds. Biodiversity evaluation and the preservation of autochthonous olive cultivars are our focus points, with the principal aim of appraisal and raising the value of Croatian monovarietal VOOs.

## 2. Materials and Methods

### 2.1. Harvesting and Olive Oil Extraction

Healthy olive fruits from four autochthonous Croatian cultivars typical for the Dalmatian region (Oblica, Lastovka, Levantinka, Krvavica) were harvested manually during the second half of October 2012 (three trees per cultivar) from genetically identified cultivars. The olive trees were all cultivated in the same experimental field of the Institute for Adriatic Crops (Kaštela, Croatia) under the same pedoclimatic and agronomic conditions. The field is located 0.5 km from the coast (43_550 N; 16_350 E) and 28 m above sea level. It is influenced by the Mediterranean climate, defined as the Csa climate type [63]. The land there is an almost flat coastal plain, and the effective soil depth is 75 cm. It is clay-loam with an alkaline reaction, with a low-to-medium level of skeleton [39]. This experimental field is 70 years old, where non-irrigated olives grow. Other cultivation practices such as plant protection, pruning, and fertilization were applied in a manner consistent with accepted commercial practices. Oblica has medium vigorous trees with a rounded canopy. Its fruits are spherical-shaped, with an average weight of 5 g and 18–21% oil yield [56]. Due to uneven ripening during the harvest period, the fruits are different colored, from green to dark purple. Lastovka has a medium lush pyramidal canopy with a short and very forked trunk. Its fruits are elliptic-shaped, with an average weight of 3 g and about 24% oil yield [56]. Lastovka has a late ripening period, and the fruit is completely black at the fully mature stage. Levantinka has a very dense rounded canopy with an elegant and smooth trunk. Its fruits are medium-sized and formed in clusters, elliptically elongated, and slightly curled towards the top. The medium fruit weight is about 4.5 g with a 20% oil content [56]. Levantinka has a medium ripening period. Krvavica has a very vigorous canopy and a strong trunk. Its fruit is spherical-shaped and slightly elongated towards the top, with an average fruit weight of 3.5 g and an 18% oil yield [56]. Krvavica has a medium ripening period, with dark purple to black colored completely ripe fruits. All fruits were harvested with a similar fruit maturity index (MI = 1.5–2.0), which is calculated using fruit skin and pulp coloration grades [26]. The fruits collected from each tree, as three biological replicates, were processed separately into oil (for each cultivar, three trees with corresponding three oil samples). The fruits were processed within 24 h from harvesting using centrifugal extraction in two phases on a laboratory small-scale Abencor system (MC2 Ingeniería y Sistemas S.L., Sevilla, Spain) that simulates industrial processing with controlled producing parameters. Approximately 1 kg of fruits was crushed using a hammer mill (MM-100) with an easily detachable stainless-steel hammer of 4.5 mm diameter. The obtained olive paste was malaxed in a thermomix bath (TB-100) for 35 min under a strictly controlled temperature ($26 \pm 1$ °C). After malaxation, the prepared olive paste was immediately subjected to centrifugation using a centrifugal machine (CF-100) with a rotation speed of 3500 rpm and a duration of 90 s. The obtained olive oil samples were clarified by additional fine centrifugation using a Hettich Universal 320R centrifuge machine (Andreas Hettich GmbH & Co. KG, Tuttlingen, Germany) with a cooling system at 18 °C and a speed of 5000 rpm for 4 min. All samples were stored in dark glass bottles under the atmosphere of nitrogen at a temperature of $-20$ °C for one month until analyses.

### 2.2. Olive Oil Qualitative Parameters

In all obtained olive oil samples, the basic quality parameters (FFA—free fatty acids, expressed % of oleic acid; PV—peroxide value, expressed as meq $O_2$/kg; and UV spectrophotometric indices at 232 and 270 nm—$K_{232}$ and $K_{270}$) were analyzed according to EEC methodology [4]. All results are expressed as mean values of three measurements. All chemicals and reagents were acquired from Carlo Erba Reagents (Val de Reuil Cedex, France).

### 2.3. Determination of Total Phenolic Content

Total phenolic content (TPC) in investigated VOOs was determined by the spectrophotometric method [64]. The separation of phenolic extracts was performed as follows: the liquid–liquid extraction of the previous prepared hexane/oil solution with a water/methanol mixture (60:40, *w/w*) was used three times in a row. The Folin-Ciocalteu reagent (Sigma-Aldrich, St. Louis, MO, USA) was used for the colorimetric reaction. The measurements of absorbances were performed in triplicate at 765 nm on the Cary 50 UV–vis spectrophotometer (Varian, CA, USA), and the results are expressed as mean values and presented as mg of Gallic acid per kg of oil.

### 2.4. Determination of Fatty Acid Composition

Fatty acid methyl esters (FAME) from VOO samples were obtained by alkaline treatment with 1M KOH in methanol using capillary column DB-WAX (film 0.25 µm; 30 m × 0.25 mm; Agilent, Santa Clara CA, USA), as described in ISO method [65]. The profiles of fatty acids in monovarietal VOOs were determined by gas chromatography separation of prepared methyl esters, according to the ISO method [66]. The results are expressed as average from three replicates for each sample.

### 2.5. Analysis of Volatile Composition by HS-SPME/GC-MS

The volatile compounds of VOOs were extracted using headspace solid-phase microextraction (HS-SPME) and analyzed by gas chromatography/mass spectrometry (GC-MS) using a Varian 3900 GC instrument coupled to a Varian Saturn 2100T ion trap mass spectrometer (Varian Inc., Harbur City, CA, USA) according to slightly modified methodology, as described by Brkić Bubola et al. [12]. Differently from the above-cited method, a VOO sample (3.5 g) was placed in a 10 mL vial. The identification of volatile compounds was performed by comparing their mass spectra with those of pure standards and to mass spectra from the NIST05 library. Additionally, the identification of 10 volatile compounds was carried out by comparing their retention times with those of pure standards. All standards had a GC purity ≥95% and were purchased from Aldrich (Steinheim, Germany) and Fluka (Buchs, Germany). Moreover, Kováts' retention indices (KIs) were determined on a polar Rtx-WAX capillary column (Restek, Bellefonte, PA, USA) by injection of a standard mixture containing the homologous series of normal alkanes (C6-C24) in pure dichloromethane and compared with the retention indices of the compounds available in the literature [18,23,38,48,67]. The relative proportions of the volatile compounds were obtained by peak area normalization. For each volatile compound, the mean proportions of three independent repetitions are reported.

### 2.6. Sensory Analyses of VOOs

A trained professional panel of the Institute for Adriatic Crops from Split, Croatia, approved by the Croatian Ministry of Agriculture, performed a sensory evaluation of monovarietal VOOs using the official IOC method [68]. All panel members conducted training according to the standard IOC method [69] using reference samples obtained directly from IOC Madrid, Spain. All oil samples were evaluated by 8 tasters. The descriptive sensory analysis was used for a detailed description of each monovarietal VOO, and positive and negative sensory attributes were perceived independently by each panel member. The IOC profile sheet used for this evaluation was expanded with positive descriptors listed in the IOC methods for sensory assessment of extra virgin olive oils with Protected Designations of Origin (PDOs) [70]. The intensity of each descriptor was graded individually by tasters using a continuous unstructured scale of 10 cm. The results are presented as the median values of the tasters' sensory perceptions [68].

### 2.7. Statistical Analyses

The results of all chemical analyses obtained with this investigation were subjected to a one-way analysis of variance (ANOVA). Firstly, the data were tested for normality and homogeneity of variance and transformed when necessary. Mean values were linked by Tukey's honest significant difference test with the level of $p \leq 0.05$. Statistical analysis was performed using Statistica v. 13.2 software (Stat-Soft Inc., Tulsa, OK, USA).

## 3. Results and Discussion

### 3.1. VOO Quality Assessment

The determination of basic quality parameters is one of the fundamental preconditions for the categorization of an olive oil. The values for FFAs, PVs, and specific coefficients of extinction at 232 and 270 nm ($K_{232}$ and $K_{270}$) and their respective $\Delta K$ values for investigated monovarietal virgin olive oil samples are presented in Table 1.

**Table 1.** Quality parameters of monovarietal virgin olive oils obtained from Dalmatian autochthonous cultivars.

| | Cultivar | | | | |
|---|---|---|---|---|---|
| Parameter | Oblica | Lastovka | Levantinka | Krvavica | EVOO * |
| FFA% (oleic acid) | $0.4 \pm 0.1$ a | $0.4 \pm 0.0$ a | $0.4 \pm 0.2$ a | $0.2 \pm 0.0$ a | $\leq 0.8$ |
| PV (meq $O_2$/kg) | $3.7 \pm 0.7$ b | $6.5 \pm 0.1$ a | $6.3 \pm 1.6$ a | $3.6 \pm 0.2$ b | $\leq 20$ |
| $K_{232}$ | $1.83 \pm 0.31$ a | $1.74 \pm 0.08$ a | $2.30 \pm 0.19$ a | $1.85 \pm 0.05$ a | $\leq 2.50$ |
| $K_{270}$ | $0.16 \pm 0.01$ a | $0.19 \pm 0.03$ a | $0.16 \pm 0.01$ a | $0.19 \pm 0.01$ a | $\leq 0.22$ |
| $\Delta K$ | $-0.00 \pm 0.00$ a | $0.00 \pm 0.00$ a | $0.00 \pm 0.00$ a | $-0.00 \pm 0.00$ a | $\leq 0.01$ |
| TPC (mg Gallic acid/kg of oil) | $438.3 \pm 12.4$ ab | $312.0 \pm 3.4$ bc | $302.1 \pm 79.0$ b | $438.9 \pm 52.0$ a | |

FFA—free fatty acid, PV—peroxide value, UV spectrophotometric indices ($K_{232}$ and $K_{270}$, $\Delta K$), TPC—total phenol content, EVOO—extra virgin olive oil. Results are expressed as mean values of three repetitions $\pm$ standard deviation. The means within each row, labeled by different letters, are significantly different (Tukey's test, $p \leq 0.05$). * Actual limits for the EVOO category [4], except for TPC.

The acidity values ranged from 0.2% to 0.4%. Peroxide value, an indicator of primary auto-oxidation products, ranged from 3.7 (Krvavica) to 6.5 (Lastovka) meq $O_2$/kg. Values of specific coefficients of extinction are measured at 232 ($K_{232}$) and 270 nm ($K_{270}$), corresponding to the maximum absorption of the conjugated dienes and trienes. They are formed during autoxidation from the hydroperoxides of unsaturated fatty acids and their fragmentation products [27]. Data for $K_{232}$ and $K_{270}$ did not show any differences among monovarietal oils. Comparing all these results with the limits set up by international regulation [4,5], it is evident that all investigated oils were in the category of extra virgin olive oil.

In addition, the values for TPC in monovarietal VOOs are presented in Table 1. Although TPC is not officially considered a quality indicator prescribed by regulation, it is a very important parameter related to the health properties of VOOs and their sensorial characteristics. The obtained results for TPC values in the four Dalmatian monovarietal olive oils indicated significant differences among the monovarietal olive oils. The highest TPC value was detected in the Krvavica oils, which did not differ from the TPC value for oils from Oblica. Oils from Lastovka and Levantinka had similar values for TPC. The mean TPC value in Oblica oil (438.3 mg/kg) were in concordance with the data obtained in the study of Jukić Špika et al. [39] for the same crop year, but still lower compared with another study of Lukić et al. [59] during the crop year of 2015, where the fruit ripening showed the strongest effect on the accumulation of phenolic compounds. Levantinka oils had a mean TPC value (302.1 mg/kg) higher than the detected value for this monovarietal oil in a previous study [61]. Additionally, the mean TPC values measured in Krvavica olive oils were much higher than those found in the literature [62]. On the contrary, in Lastovka oils, the mean TPC value (312.0 mg/kg) was slightly lower compared to the highest TPC values reported in previous research [61]. These statements could be related to different maturity indices of fruits and also to different technological processes since our samples were processed in a pilot-scale system. It is known that total phenolics content changes during the ripening period and depends on the crop year and also the cultivar [7,8,29,39]. In our study, the olive fruits of each cultivar had the same maturity index at the moment of harvesting (MI = 1.5–2.0) in order to decrease the possible influence of this factor on TPC in the analyzed olive oil samples.

### 3.2. Fatty Acid Composition

Monovarietal olive oils differ in the fatty acids' composition depending on the olive cultivar [50]. Different authors have studied the composition of fatty acids in order to characterize the olive cultivar [47,56,60]. The fatty acid composition of Dalmatian monovarietal olive oils is presented in Table 2. In all investigated oil samples, palmitic acid (C16:0) has been identified with a medium value of 12.58%. Palmitic acid (C16:0), the major saturated fatty acid in olive oil, had the highest proportion in Krvavica olive oils (about 14%), while the lowest content was detected in Levantinka oils (about 12%) even though no statistically significant difference in palmitic acid content among studied cultivars was determined. The significant highest value of palmitoleic acid (C16:1) was detected in Krvavica oils. Both heptadecanoic and heptadecenoic acids (C17:0 and C17:1, n-9) were detected in low amounts in all oil samples, with an average of 0.05% and 0.11%, respectively. Regarding heptadecanoic acid, the highest content was detected in Lastovka oils. Comparing our results with other studies, Italian cultivar Frantoio [71] had higher C17:1 values (0.20–0.30%) measured for different growing sites.

**Table 2.** Fatty acid composition of monovarietal virgin olive oils obtained from Dalmatian autochthonous cultivars.

| Fatty Acid * (% of Total) | Cultivars | | | | EVOO * |
|---|---|---|---|---|---|
| | Oblica | Lastovka | Levantinka | Krvavica | |
| C16:0 | 12.26 ± 0.38 a | 12.32 ± 0.59 a | 11.72 ± 0.13 a | 14.00 ± 0.93 a | 7.50–20.00 |
| C16:1 | 0.75 ± 0.06 b | 0.75 ± 0.04 b | 1.04 ± 0.24 b | 2.40 ± 0.43 a | 0.30–3.50 |
| C17:0 | 0.03 ± 0.01 b | 0.10 ± 0.01 a | 0.04 ± 0.00 b | 0.04 ± 0.00 b | ≤0.40 |
| C17:1, n-9 | 0.08 ± 0.04 a | 0.19 ± 0.06 a | 0.08 ± 0.01 a | 0.07 ± 0.00 a | ≤0.60 |
| C18:0 | 1.43 ± 0.51 a | 1.71 ± 0.48 a | 1.90 ± 0.10 a | 1.40 ± 0.74 a | 0.50–5.00 |
| C18:1 | 75.91 ± 1.41 a | 71.81 ± 1.15 b | 76.17 ± 0.30 a | 77.14 ± 0.34 a | 55.00–83.00 |
| C18:2 | 9.66 ± 0.85 b | 11.63 ± 0.29 a | 6.78 ± 0.19 c | 4.62 ± 0.04 d | 2.50–21.00 |
| C18:3 | 0.66 ± 0.07 a | 0.66 ± 0.04 a | 0.66 ± 0.03 a | 0.75 ± 0.15 a | ≤1.00 |
| C20:0 | 0.39 ± 0.06 a | 0.48 ± 0.09 a | 0.53 ± 0.11 a | 0.30 ± 0.00 a | ≤0.60 |
| C20:1, n-9 | 0.36 ± 0.03 a | 0.38 ± 0.11 a | 0.40 ± 0.08 a | 0.41 ± 0.00 a | ≤0.50 |
| C22:0 | 0.11 ± 0.01 a | 0.10 ± 0.01 a | 0.11 ± 0.01 a | 0.10 ± 0.00 a | ≤0.30 |
| C24:0 | 0.10 ± 0.01 a | 0.10 ± 0.00 a | 0.10 ± 0.00 a | 0.10 ± 0.00 a | ≤0.20 |
| C18:1/C18:2 | 7.92 ± 0.49 c | 6.18 ± 0.09 d | 11.24 ± 0.30 b | 16.83 ± 0.53 a | - |
| MUFA | 77.10 ± 1.47 ab | 73.12 ± 1.35 b | 77.68 ± 0.34 a | 80.01 ± 0.76 a | - |
| PUFA | 10.31 ± 0.92 a | 12.28 ± 0.35 a | 7.44 ± 0.21 b | 5.36 ± 0.18 c | - |
| UFA | 86.75 ± 2.32 a | 84.74 ± 1.65 a | 84.46 ± 0.53 a | 84.63 ± 0.79 a | - |
| SFA | 14.33 ± 0.85 a | 14.79 ± 0.73 a | 14.39 ± 0.33 a | 15.94 ± 1.66 a | - |
| MUFA/PUFA | 7.50 ± 0.52 c | 5.95 ± 0.04 d | 10.43 ± 0.25 b | 14.91 ± 0.36 a | - |

* C16:0—palmitic acid; C16:1—palmitoleic acid; C17:0—heptadecanoic acid; C17:1—heptadecenoic acid; C18:0—stearic acid; C18:1—oleic acid; C18:2—linoleic acid; C18:3—linolenic acid; C 20:0—arachidonic acid; C20:1—gadoleic acid; C22:0—behenic acid; C24:0—lignoceric acid. MUFA—monounsaturated fatty acid; PUFA—polyunsaturated fatty acid; UFA—unsaturated fatty acid; SFA—saturated fatty acid. Results are expressed as mean values of three repetitions ± standard deviation (SD). * Actual limits for extra virgin olive oil category [4]. The means within each row, labeled by different letters, are significantly different (Tukey's test, $p \leq 0.05$).

For all analyzed oils, the content of stearic acid (C18:0) had a medium value of 1.61%, with no significant differences detected among cultivars. All analyzed oil samples had contents of monounsaturated oleic acid (C18:1) higher than 70%. The significant difference for C18:1 was observed in Lastovka oils, which had the lowest value. At the same time, this oil had the highest content of linoleic (C18:2) compared to other studied monovarietal oils. In Oblica VOOs, the mean value of C18:2 was 9.66%, followed by Levantinka with 6.78%, while the significantly lowest content of linoleic fatty acid was detected in Krvavica olive oils (4.62%). In comparison with data from another study [39], Oblica had a high C16:0 content (average 13.43%) as well as a medium content of C18:2 (average 11.22%). C18:2 content in the Leccino cultivar grown in Croatia [39] was of an average of 6.99%

(5.53–9.74%), while Frantoio oils had C18:2 contents in the range of 6.20—8.20% (average 7.38%) [71].

There was no significant difference among other less represented fatty acids (arachidonic, gadoleic, behenic, and lignoceric acid) identified in the obtained VOOs (Table 2). Based on EU regulation [4], all analyzed monovarietal olive oils were placed in the category of extra virgin olive oil, according to fatty acid composition. It was evident that the Krvavica cultivar showed clear differences in fatty acid composition compared to oils from other cultivars. Therefore, this feature could be considered a characterization point for Krvavica olive oils, as it was highlighted in another study that fatty acid composition can be used as a base for the characterization and evaluation of VOOs [62]. Different studies [40,61,72] have investigated the association between oleic and linoleic acids (18:1/18:2) and the oxidative stability of VOO. The results showed that the ratio of monounsaturated/polyunsaturated fatty acids is one of the key factors responsible for evaluating the oxidative stability of VOOs, and it is used as a parameter for oil characterization [73]. In our study, a significant difference for the C18:1/C18:2 ratio was observed among studied cultivars. Oblica and Lastovka oils showed a stable index expressed by the ratio of oleic acid to linoleic acid, with values near or higher than 7, while Levantinka and Krvavica oils had higher medium ratios of oleic acid/linoleic acid (11.24% and 16.83%, respectively), which implicate their high stability and resistance to oxidative degradation and, consequently, longer shelf life under correct storage conditions. The same conclusion was confirmed by the ratios of MUFA/PUFA (Table 2). Since VOO consists mainly of monounsaturated fatty acid (C18:1, MUFA) and significant amounts of polyunsaturated fatty acids (C18:2, C18:3, PUFA), it is considered a unique oil among other vegetable oils. There is an EFSA-approved health claim [74] on the unsaturated fatty acids, saying: 'Replacing saturated fats in the diet with unsaturated fats contributes to the maintenance of normal blood cholesterol levels'. The claim may be used only for food that is high in unsaturated fatty acids, and VOO is, for sure, one of them.

### 3.3. Volatile Profiling of Dalmatian Monovarietal VOOs

The results of the volatile composition of the analyzed monovarietal olive oils from the four Dalmatian autochthonous cultivars are presented in Table 3. In total, 21 volatile compounds were detected, mainly aldehydes, alcohols, esters, terpenes, and organic acids. The same number of volatile compounds (21) was found in south Italian extra virgin olive oils in the study of Giuffrè et al. [43]. Different origin olive oils were studied by Luna et al. [50], and they found that Spanish and Italian VOOs had the highest values of total volatiles, with concentrations ranging from 19.0 to 27.0 mg/kg. Coratina was the variety with the highest value and Frantoio with the lowest one. The Spanish oils showed the widest concentration range (9.83–32.9 mg/kg), Nevado Azul being the variety with the highest value and Hojiblanca with the lowest one. With respect to Greek virgin olive oils, their concentrations ranged from 10.7 to 21.2 mg/kg. Finally, the non-European samples showed a total content of volatiles below 20 mg/kg, with the exception of Chemlal of Kabylie [50]. In our paper, even though no significant differences between monovarietal oils in total aldehydes, alcohols, esters, terpenes, and organic acids, as well as total C5 and C6 volatiles, were detected (Figure 1), some particularities in the volatile compound composition of monovarietal olive oils were found (Table 3).

**Table 3.** Volatile composition of virgin olive oils obtained from Dalmatian autochthonous cultivars (identification and mean proportion values of peak area) detected by HS-SPME/GC-MS.

| Compound | Identification Method | KI * | KI-Ref | Oblica Mean * (%) ±SD | Lastovka Mean (%) ±SD | Levantinka Mean (%) ±SD | Krvavica Mean (%) ±SD |
|---|---|---|---|---|---|---|---|
| *Aldehydes* | | | | | | | |
| Hexanal | KI, MS, RT | 1072 | 1074 [1], 1086 [2], 1073 [3] | 12.74 ± 4.59 b | 23.69 ± 2.87 a | 1.34 ± 1.05 c | 13.76 ± 0.04 b |
| Z-3-Hexenal | KI, MS | 1135 | 1137 [1], 1115 [2] | 2.87 ± 3.55 a | 0.48 ± 0.01 a | 0.25 ± 0.28 a | 5.91 ± 1.61 a |
| Heptanal | KI, MS | 1191 | 1184 [1], 1190 [2] | 0.04 ± 0.02 b | 0.39 ± 0.03 a | 0.03 ± 0.02 b | 0.16 ± 0.07 b |
| E-2-Hexenal | KI, MS, RT | 1208 | 1216 [1], 1225 [2], 1129 [3] | 48.03 ± 9.66 a | 43.3 ± 16.92 a | 74.84 ± 1.48 b | 45.28 ± 1.44 a |
| (E,E) or (E,Z)-2,4-Hexadienal | KI, MS | 1384 | 1397 [1], 1402 [3] | 1.45 ± 1.42 a | 0.26 ± 0.02 a | 0.30 ± 0.35 a | 1.55 ± 1.83 a |
| (E,E) or (E,Z)-2,4-Hexadienal | KI, MS | 1388 | 1397 [1], 1402 [3] | 8.23 ± 8.48 a | 4.15 ± 0.5 a | 1.84 ± 2.45 a | 13.36 ± 2.67 a |
| (Z)-2-Heptenal | KI, MS | 1314 | 1320 [4] | 0.08 ± 0.07 b | 1.08 ± 0.1 a | 0.10 ± 0.04 b | 0.18 ± 0.1 b |
| Octanal | KI, MS, RT | 1282 | 1288 [1], 1296 [2], 1297 [3] | 0.05 ± 0.02 a | 7.50 ± 8.72 a | 0.05 ± 0.02 a | 0.22 ± 0.11 a |
| E,E-2,4- Heptadienal | KI, MS | 1451 | 1463 [1] | 0.18 ± 0.18 a | 0.44 ± 0.27 a | 0.22 ± 0.23 a | 0.54 ± 0.03 a |
| *Total aldehydes* | | | | 73.67 ± 8.68 a | 81.29 ± 6.19 a | 78.97 ± 0.85 a | 80.94 ± 4.32 a |
| *Alcohols* | | | | | | | |
| 1-Penten-3-ol | KI, MS | 1159 | 1164 [1], 1166 [2], 1163 [3] | 1.61 ± 1.26 a | 2.42 ± 2.55 a | 0.51 ± 0.16 a | 2.12 ± 0.17 a |
| Z-2-Penten-1-ol | KI, MS, RT | 1302 | 1320 [1], 1329 [2], 1321 [3] | 0.90 ± 0.88 a | 2.44 ± 0.61 a | 0.73 ± 0.41 a | 1.88 ± 2.21 a |
| E-2-Penten-1-ol | KI, MS | 1310 | 1320 [2], 1333 [3] | 2.40 ± 1.48 b | 6.22 ± 0.06 a | 0.84 ± 0.38 b | 2.74 ± 0.41 b |
| Hexanol | KI, MS, RT | 1344 | 1357 [1], 1362 [2], 1360 [3], 1354 [4] | 0.20 ± 0.10 b | 1.01 ± 0.19 b | 4.49 ± 1.12 a | 0.40 ± 0.09 b |
| E-3-Hexen-1-ol | KI, MS, RT | 1354 | 1366 [1], 1372 [2], 1372 [3] | 0.15 ± 0.01 a | 0.04 ± 0 b | 0.20 ± 0.05 a | 0.10 ± 0.01 ab |
| Z-3-Hexen-1-ol | KI, MS, RT | 1374 | 1385 [1], 1392 [2], 1385 [3], 1388 [4] | 13.25 ± 6.33 a | 1.24 ± 0.27 a | 7.35 ± 0.98 a | 2.97 ± 1.37 a |
| *Total alcohols* | | | | 18.51 ± 2.62 a | 13.37 ± 3.19 a | 14.12 ± 1.91 a | 10.21 ± 0.16 a |
| *Esters* | | | | | | | |
| Methyl acetate | KI, MS | <1000 | 800 [1] | 0.4 ± 0.07 a | 0.34 ± 0.02 a | 0.29 ± 0.26 a | 0.27 ± 0.04 a |
| Ethyl acetate | KI, MS, RT | <1000 | 892 [1], 895 [3] | 0.06 ± 0.02 a | 0.78 ± 0.79 a | 0.49 ± 0.52 a | 0.11 ± 0.01 a |
| Hexyl acetate | KI, MS, RT | 1268 | 1247 [1], 1281 [2], 1269 [4] | 0.00 ± 0.00 b | 0.63 ± 0.02 a | 0.08 ± 0.08 b | 0.05 ± 0.00 b |
| *Total esters* | | | | 0.46 ± 0.05 a | 1.75 ± 0.80 a | 0.86 ± 0.70 a | 0.43 ± 0.03 a |
| *Terpenes* | | | | | | | |
| α-Copaene | KI, MS | 1487 | 1481 [1], 1500 [2], 1505 [3] | 1.30 ± 1.53 a | 0.34 ± 0.17 a | 2.10 ± 1.58a | 3.15 ± 1.54 a |
| *Total terpenes* | | | | 1.30 ± 1.53 a | 0.34 ± 0.17 a | 2.10 ± 1.58a | 3.15 ± 1.54 a |
| *Organic acids* | | | | | | | |
| Acetic acid | KI, MS, RT | 1430 | 1448 [1] | 1.17 ± 1.18 a | 2.67 ± 3.66 a | 0.03 ± 0.03 a | 4.84 ± 5.59 a |
| Propanoic acid | KI, MS | 1519 | 1528 [1] | 0.06 ± 0.04 a | 0.14 ± 0.04 a | 0.13 ± 0.11 a | 0.27 ± 0.04 a |
| *Total organic acids* | | | | 1.23 ± 1.14 a | 2.81 ± 3.62 a | 0.16 ± 0.14 a | 5.11 ± 5.64 a |

* Results are expressed as mean values of three independent repetitions ± standard deviation. The means within each row, labeled by different letters, are significantly different (Tukey's test, $p \leq 0.05$). Identification methods: RT—identification by comparison with retention times and mass spectra of pure standards; MS—identification by comparison with mass spectra from the NIST05 library; KI—identification by comparison with Kováts' retention indexes from the literature (KIref) ([1] [38]; [2] [68]; [3] [16]; [4] [48]). KI*—Kováts' retention indexes on Rtx-WAX capillary columns.

In general, the most prevalent volatile compound in all oil samples was C6 aldehyde *E*-2-hexenal (Table 3), which contributes to green, fruity, bitter, and astringent sensory characteristics [48] and also bitter almond and green grass notes [51]. The same findings were presented in a study by Giuffrè et al. [43], where E−2-hexenal was found at an average of 28.3% in all analyzed south Italian extra virgin olive oils. The highest value for *E*-2-hexenal in our study was detected in Levantinka oils (74.84% of total peak area), while the other three monovarietal olive oils had similar values of *E*-2-hexenal. Significant differences among cultivars were found in certain aldehydes, namely, hexanal, heptanal, and *Z*-2-heptenal (Table 3). Lastovka oils had the highest heptanal and *Z*-2-heptenal amounts. The same monovarietal oil had the highest amount of hexanal (Table 3), C6 aldehyde is associated with the sensory characteristics of green fruitiness, apple, and green grass [48], while the other cultivars had a lower amount. Levantinka oil had the lowest amount of hexanal (1.34% of total peak area). In our study, Krvavica oils had a higher hexanal content than those found in an earlier investigation [62], probably due to a higher maturity index (MI = 4.3) than in our study (MI = 1.5–2). Hexanal, *E*-2-hexenal, and *Z*-3-hexenal are C6 aldehydes that arise during the malaxation of olive paste, and they occur in a series of enzymatic reactions known as the lipoxygenase pathway (LOX), from linoleic or alpha linolenic acids with lipoxygenase-catalyzed oxidation [50]. Hydroperoxide lyase is quite dependent on the variety, so its activity and concentration lead to different volatile profiles and, thus, the sensory properties of monovarietal oils [17], as confirmed in literature [59].

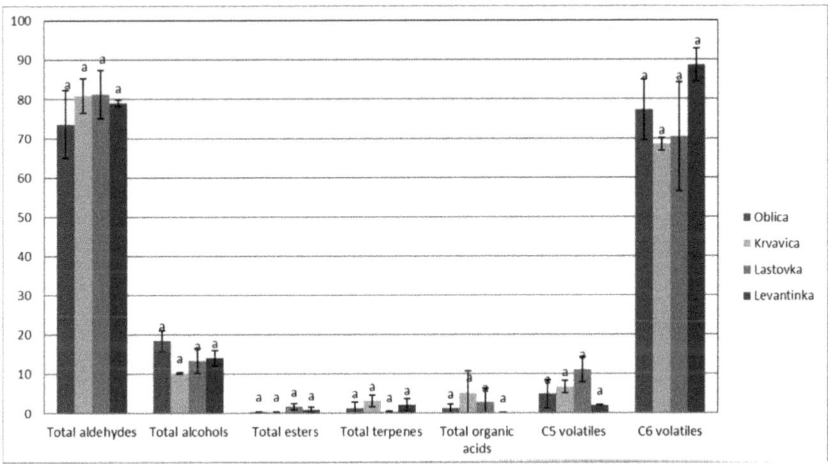


**Figure 1.** The amounts of volatile compounds (percentage of total peak area) in VOOs from Dalmatian autochthonous cultivars (results are expressed as means of 3 independent repetitions). The means labeled by different letters are significantly different (Tukey's test, $p \leq 0.05$).

The C5 and C6 volatile contents mainly depend on genetic origin, which is responsible for enzymatic expression, and agronomic and technological parameters that strongly influence enzymatic activity [10]. The same group of volatile compounds are also present in different fruits and vegetables [50]; thus, it is understood why some olfactory sensations in VOOs are described as fruity sensations of apple, banana, almond, or artichoke.

Alcohols in VOOs are associated with positive sensory characteristics, such as green, fruity, bitter, and aromatic, although they have weaker sensory importance than aldehydes [48]. The highest, although not significantly, total alcohols were detected in Oblica oils (Table 3), mainly due to the highest values of *Z*-3-hexen-1-ol (13.25% of total peak area) (Table 3), which is related to the banana fruity sensation [37,50]. These results support the banana fruitiness of Oblica VOO as a cultivar descriptor, as declared in a previous paper [49] and confirmed in our study as well (Figure 1). Other alcohols were found to be prevalent in Levantinka (hexanol) and Lastovka (*E*-2-penten-1-ol) VOOs.

Esters detected in analyzed olive oils had low values in general (Table 3). This group of compounds is related to sweet and fruity sensory attributes [37,55], and significant differences among cultivars were not found except in the case of hexyl acetate. The highest value for this ester was measured in Lastovka oils (0.63% of total peak area), while it was not detected in Oblica oils. C6 esters have synergistic interactions with other aromatic compounds [46].

Regarding the terpenes, only one was detected in Dalmatian olive oils: α-copaene. No significant difference was found among monovarietal oils in α-copaene (Table 3), but in Krvavica oils, higher values were detected compared to data in a previous study [62].

The presence of acetic and propanoic acids indicates the possibility of microbial fruit fermentation or some other incorrect step during fruit handling and is linked to sour and pungent attributes [37,50]. These acids also can generate negative sensory defects of winey–vinegary taste in VOOs [50]. In our study, total organic acids had very low values in all analyzed olive oils (Table 3), which supported the results of the sensorial analysis, where no negative attributes were detected (Figure 2).

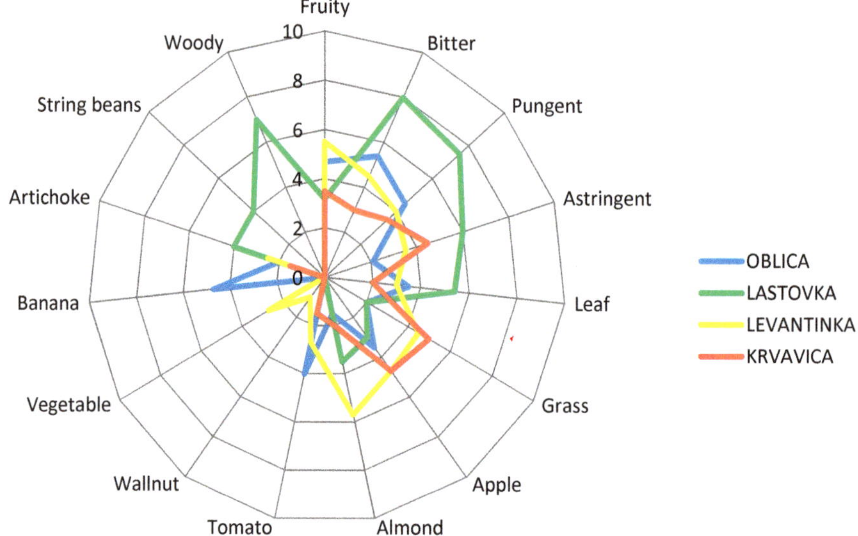


**Figure 2.** Sensory profiles of virgin olive oils obtained from four Dalmatian autochthonous cultivars. Results are expressed as the mean value of medians of three repetitions for each descriptor.

### 3.4. Sensory Characteristics of Monovarietal VOOs

Sensory analyses of VOOs are a crucial step in the determination of the olive oil category, according to EU legislation [4,68]. Currently, sensory assessment by a well-trained panel group of experts remains an inevitable instrument in the organoleptic appraisal and aroma recognition of VOOs. During sensory analyses of VOOs, according to IOC methodology [68], the positive and negative attributes are perceived by smelling and tasting the olive oil sample. Besides the fruitiness that is inevitable for the category of VOO, bitterness and pungency are very desirable features of VOO, although those characteristics are not reflected as important in commodity classification [75]. The attributes of bitter and spicy oil are due to the presence of phenolic compounds. The bitter taste is attributed to compounds of the aglycone and dialdehydic forms of decarboxymethyl oleuropein and other forms of oleuropein aglycone. The pungent note has been attributed to the aglycone form of the dialdehydic of decarboxymethyl ligstroside [37]. In order to provide a more detailed organoleptic description and create a sensory profile of studied monovarietal olive oils, quantitative descriptive sensory analysis was carried out by a professional sensory panel using the IOC methodology developed for sensory evaluation of olive oils with protected designations of origin (PDOs) [70]. The sensory descriptive profile of monovarietal VOOs from our study is presented in Figure 2. Oblica olive oils were characterized with typical fruity sensations of banana, tomato, and apple, with well-pronounced and equilibrated bitterness and pungency (Figure 2). Distinct banana fruitiness was detected only in Oblica oils. Astringency and an almond sensation had the lowest intensity in Oblica olive oils compared to oils from other olive varieties. A slight note of green grass and artichoke fruitiness was also perceived in Oblica oils. A grass sensation was of equal intensity in Oblica and Lastovka oils and lower by 2.25 times than its value in Levantinka and 2.5 times than in Krvavica oils. Among average consumers, Oblica is considered to be a rather sweet olive oil [61]. However, it is confirmed that in the right maturity moment, its oil has satisfactory and rounded bitterness and pungency [59], two positive attributes that are linked to the positive health impact of this oil due to the presence of phenolic compounds [76]. The sensory profile of Lastovka oils showed the highest astringency intensity compared to other studied cultivars: 3 times higher than in Oblica oils, 1.7 times higher than in Levantinka oils, and 1.33 times higher than in Krvavica oils. Lastovka olive oil was characterized by mild fruitiness, similar to green leaves and apple. No tomato fruitiness was detected, unlike oils from the other cultivars. In Lastovka olive oil, high intensity of a woody sensation was perceived (which could be explained by the high amount of hexanal found in these oils) (Table 3). Although the presence of hexanal is regularly linked to green/grassy aromas in olive oil [50], Mayuoni-Kirshinbaum and Porat [77] reported the possible connection between hexane and woody flavor in their review of the flavor of pomegranate fruit. A woody attribute was perceived only in Lastovka oils, the same as a string bean sensation. Literature findings have declared that certain volatile compounds, mainly from the sesquiterpenes group, could also be responsible for the woody sensory sensation in fruit [78]. This fact was confirmed by Lukić et al. [78] in the case of Lastovka olive oil, applying a combined targeted and untargeted profiling of volatile aroma compounds. There is a possibility that the woody aroma in Lastovka oils appeared due to the presence of hexanal [77] or some sesquiterpenes [78]. The almond fruitiness found in Lastovka oils could be linked to the presence of $Z$-2-penten-1-ol, which is positively related to this sensory note [78].

In Lastovka oils, which are characterized by the strongest bitterness, pungency, and astringency compared to olive oils obtained from other cultivars, the high value of ethyl acetate and 1-penten-3-ol (Table 3) could have an influence on the taste characteristics since these volatiles are correlated with the 'taste' notes of pungency and astringency [2,37]. The presence of ethyl acetate and 1-penten-3-ol may have contributed to the astringency in Lastovka oils, in addition to the phenolics found in the oil, although not present in high amounts. Research was conducted on regular Italian consumers of olive oil who were familiar with olive oil on a daily basis, and the taste–smell interaction between bitterness and green grass odors was investigated [79]. The results confirmed that the green odor note had a significant positive effect on consumers' perception of bitterness, which increased in the presence of green grass odors [79]. In our study, that could be the case for Levantinka oils, where the highest amount of $E$-2-hexenal was detected (Table 3), which could enhance the bitterness sensation of this oil. Additionally, the bitterness and grass sensations were equally graded in this oil (Figure 2). Moreover, Levantinka oils were characterized with the highest level of almond fruitiness compared with olive oils from the other three studied cultivars, accompanied with apple nuances, which was confirmed by the highest value for $E$-2-hexenal among all investigated monovarietal oils (Table 3), while tomato and vegetable notes were less graded. A slight note of walnut was detected only in Levantinka olive oils. This oil had very well-balanced bitterness and pungency, which could be related to the presence of a green grass odor, as stated in literature [79]. Mild to medium fruitiness with grass and ripe apple notes were the olfactory characteristics of Krvavica VOOs. This cultivar could be a preferable choice for the majority of consumers who choose milder and sweeter olive oils compared to bitter and pungent ones [53]. Compared to other olive oils in this study, Krvavica oils had the highest level of grass fruitiness and the lowest level of artichoke attributes. The bitterness and pungency in Krvavica oils had the lowest intensity among all studied cultivars but were still delicate and well balanced, in addition to astringency, although the TPC content in this oil had a relatively high mean value (Table 1).

Caporale et al. [79] classified olive oil bitterness according to TPC values into four categories: unnoticeable bitterness (equal or lower than 220 mg/kg), slight bitterness (220–340 mg/kg), bitter oils (340–410 mg/kg), and quite bitter or very bitter oils (higher than 410 mg/kg). Considering this grouping, it can be concluded that in our investigation, two monovarietal VOOs can be placed in the second group as slight bitter oils (Lastovka and Levantinka), and two other oils (Oblica and Krvavica) belong in the fourth group as quite bitter oils. The sensory profiles of the studied monovarietal Dalmatian olive oils (Figure 2) do not follow this grouping in full. VOO aroma is the effect of complex reactions inside the product (enzymatic, chemical, and so forth) that lead to the formation of volatile compounds [1]. Positive and negative synergisms can appear, and this could result in new types of perceptions and interactions between taste and odor [1]. Considering this claim, it can be concluded that besides the phenolics present in olive oils, volatile compounds play an important role or have a synergistic effect on certain sensory attributes of olive oil. The phenolic compounds were attested to have an important role in the intensity and timing of the release of certain aroma compounds during the consumption of virgin olive oil. During sensory analyses, high levels of VOO phenolic compounds resulted in a smaller total release of 1-penten-3-one, E-2-hexenal, and esters during the swallowing of the olive oil sample [80]. This fact could be explained by the formation of complexes between phenolic compounds and salivary proteins that capture aroma compounds and, therefore, decrease their volatility during the organoleptic assessment of olive oil. Since phenolic compounds can affect the release of VOO aroma compounds during its consumption, thereby influencing flavor perception and consumer acceptance, this interesting link should be further investigated in the case of monovarietal Croatian olive oils.

## 4. Conclusions

The natural variation and great variability of VOO volatile compounds offer a wide range of different flavor and aroma characteristics. Autochthonous olive cultivars provide important raw materials for the production of specific monovarietal olive oils that have the unique opportunity to gain high market positioning and obtain higher prices. In this paper, the chemical composition and sensory characteristics of Dalmatian monovarietal VOOs from a small-scale plant were compared and their peculiar characteristics demonstrated; on this basis, they can find their place in the demanding olive oil world market. Oblica olive oils, with typical banana fruitiness (present only in this cultivar oil and confirmed with the highest value of Z-3-hexen-1-ol) and apple and tomato notes, had clear and equilibrated bitterness and pungency. In this oil, a slight sensation of green grass and artichoke fruitiness was also detected. Olive oils from Lastovka had the highest amount of several volatiles (hexanal, heptanal, and Z-2-heptenal). Sensory attributes that confirmed this are mild fruitiness that recalls green leaves, almond, and apple, with a high intensity of a unique woody sensation, strong bitterness, pungency, and the highest astringency level among all studied cultivars. Levantinka olive oils had the highest amount of C6 aldehyde E-2-hexenal, the most prevalent volatile compound in all investigated oil samples. This oil had the highest level of almond fruitiness, with perceived apple notes, very well-balanced bitterness, and pungency. Krvavica olive oils differed from the analyzed monovarietal VOOs, particularly by fatty acid composition (highest proportion of palmitic acid, significantly highest value of palmitoleic acid, significantly lowest content of linoleic acid, and a high ratio between monounsaturated and polyunsaturated fatty acids). Olive oils from Krvavica had the highest level of grass fruitiness compared to other studied olive cultivars. In these oils, delicate bitterness and pungency were well balanced with astringency. The obtained results in this study contributed to the recognition of the specificity of Croatian monovarietal olive oils, which could contribute to the protection and preservation of the national Croatian olive genetic pool. The results of this study are also important data for producers to encourage them to create their own unique high-quality VOO based on domestic autochthonous olives.

**Author Contributions:** Conceptualization, M.Ž. and K.B.B.; Methodology, M.Ž.; Formal analysis, M.Ž., M.J.Š. and K.B.B.; Resources: M.Ž. and M.J.Š.; Data curation, M.Ž., M.M.O. and K.B.B.; Data analyses: M.J.Š. and K.B.B.; writing—original draft preparation, M.Ž.; Writing—review and editing, M.Ž., M.J.Š., M.M.O. and K.B.B.; supervision, M.Ž. All authors have read and agreed to the published version of the manuscript.

**Funding:** This research was funded by the project Biodiversity and Molecular Plant Breeding, Centre of Excellence for Biodiversity and Molecular Plant Breeding (CoE CroP-BioDiv), Zagreb, Croatia (grant number KK.01.1.1.01.0005), and partially funded by Split-Dalmatia County.

**Institutional Review Board Statement:** Not applicable.

**Informed Consent Statement:** Not applicable.

**Data Availability Statement:** Not applicable.

**Acknowledgments:** The authors wish to thank Blanka Anđelić for technical assistance and all panel members who participated in the sensory evaluation of VOOs.

**Conflicts of Interest:** The authors declare no conflict of interest.

## References

1. Campestre, C.; Angelini, G.; Gasbarri, C.; Angerosa, F. The compounds responsible for the sensory profile in monovarietal virgin olive oils. *Molecules* **2017**, *22*, 1833. [CrossRef]
2. Pouliarekou, E.; Badeka, A.; Tasioula-Margari, M.; Kontakos, S.; Longobardi, F.; Kontominas, M.G. Characterization and classification of Western Greek olive oils according to cultivar and geographical origin based on volatile compounds. *J. Chromatogr. A* **2011**, *1218*, 7534–7542. [CrossRef]
3. Foscolou, A.; Critselis, E.; Panagiotakos, D. Olive oil consumption and human health: A narrative review. *Maturitas* **2018**, *118*, 60–66. [CrossRef]
4. European Union Commission. Commission Delegated Regulation No 2016/2095 of 26 September 2016 amending Regulation (EEC) No 2568/91 on the characteristics of olive oil and olive-residue oil and on the relevant methods of analysis. *J. Eur. Union* **2016**, *L326*, 1–6.
5. International Olive Council. *Trade Standard on Olive Oils and Olive Pomace Oil*; COI/T.15/NC No 3/Rev.1; International Olive Council: Madrid, Spain, 2021.
6. Veneziani, G.; Esposto, S.; Taticchi, A.; Urbani, S.; Selvaggini, R.; Sordini, B.; Servili, M. Characterization of phenolic and volatile composition of extra virgin olive oil extracted from six Italian cultivars using a cooling treatment of olive paste. *LWT Food Sci. Technol.* **2018**, *87*, 523–528. [CrossRef]
7. Romero, N.; Saavedra, J.; Tapia, F.; Sepúlveda, B.; Aparicio, R. Influence of agroclimatic parameters on phenolic and volatile compounds of Chilean virgin olive oils and characterization based on geographical origin, cultivar and ripening stage. *J. Sci. Food Agric.* **2016**, *96*, 583–592. [CrossRef]
8. Tura, D.; Failla, O.; Bassi, D.; Pedò, S.; Serraiocco, A. Cultivar influence on virgin olive (*Olea europea* L.) oil flavor based on aromatic compounds and sensorial profile. *Sci. Hortic.* **2008**, *118*, 139–148. [CrossRef]
9. Angerosa, F.; Servili, M.; Selvaggini, R.; Taticchi, A.; Esposto, S.; Montedoro, G.F. Volatile compounds in virgin olive oil: Occurrence and their relationship with the quality. *J. Chromatogr. A* **2004**, *1054*, 17–31. [CrossRef]
10. Tura, D.; Prenzler, P.D.; Bedgood, D.R.; Antolovich, M.; Robards, K. Varietal and processing effects on the volatile profile of Australian olive oils. *Food Chem.* **2004**, *84*, 341–349. [CrossRef]
11. Brkić Bubola, K.; Koprivnjak, O.; Sladonja, B.; Lukić, I. Volatile compounds and sensory profiles of monovarietal virgin olive oils from Buža, Črna and Rosinjola cultivars in Istria (Croatia). *Food Technol. Biotechnol.* **2012**, *50*, 192–198.
12. Brkić Bubola, K.; Krapac, M.; Lukić, I.; Sladonja, B.; Autino, A.; Cantini, C.; Poljuha, D. Morphological and molecular characterization of Bova olive cultivar and aroma fingerprint of its oil. *Food Technol. Biotechnol.* **2014**, *52*, 342–350.
13. Benincasa, C.; De Nino, A.; Lombardo, N.; Perri, E.; Sindona, G.; Tagarelli, A. Assay of aroma active components of virgin olive oils from southern Italian regions by SPME-GC/ion trap mass spectrometry. *J. Agric. Food Chem.* **2003**, *51*, 733–741. [CrossRef]
14. Angerosa, F.; Mostallino, R.; Basti, C.; Vito, R. Virgin olive oil odour notes. Their relationship with volatile compounds from lipoxygenase pathway and secoiridoid compounds. *Food Chem.* **2000**, *68*, 283–287. [CrossRef]
15. Sánchez-Ortiz, A.; Bejaoui, M.A.; Quintero-Flores, A.; Jiménez, A.; Beltrán, G. Biosynthesis of volatile compounds by hydroperoxide lyase enzymatic activity during virgin olive oil extraction process. *Food Res. Int.* **2018**, *111*, 220–228. [CrossRef] [PubMed]
16. Sánchez-Ortiz, A.; Pérez, A.G.; Sanz, C. Synthesis of aroma compounds of virgin olive oil: Significance of the cleavage of polyunsaturated fatty acid hydroperoxides during the oil extraction process. *Food Res. Int.* **2013**, *54*, 1972–1978. [CrossRef]

17. Pizarro, C.; Rodríguez-Tecedor, S.; Pérez-del-Notario, N.; González-Sáiz, J.M. Recognition of volatile compounds as markers in geographical discrimination of Spanish extra virgin olive oils by chemometric analysis of non-specific chromatography volatile profiles. *J. Chromatogr. A* **2011**, *1218*, 518–523. [CrossRef]

18. Kandylis, P.; Vekiari, A.S.; Kanellaki, M.; Grati Kamoun, N.; Msallem, M.; Kourkoutas, Y. Comparative study of extra virgin olive oil flavor profile of Koroneiki variety (Olea europaea var. Microcarpa alba) cultivated in Greece and Tunisia during one period of harvesting. *Lebensm. Wiss. Technol.* **2011**, *44*, 1333–1341. [CrossRef]

19. Kritioti, A.; Paikousis, L.; Drouza, C. Characterization of the volatile profile of virgin olive oils of Koroneiki and Cypriot cultivars, and classification according to the variety, geographical region and altitude. *LWT* **2020**, *129*, 109543. [CrossRef]

20. Manai, H.; Mahjoub-Haddada, F.; Oueslati, I.; Daoud, D.; Zarrouk, M. Characterization of monovarietal virgin olive oils from six crossing varieties. *Sci. Hortic.* **2008**, *115*, 252–260. [CrossRef]

21. Peres, F.; Jeleń, H.H.; Majcher, M.M.; Arraias, M.; Martins, L.L.; Ferreira-Dias, S. Characterization of aroma compounds in Portuguese extra virgin olive oils from Galega Vulgar and Cobrançosa cultivars using GC–O and GC × GC–ToFMS. *Food Res. Int.* **2013**, *54*, 1979–1986. [CrossRef]

22. Bajoub, A.; Sánchez-Ortiz, A.; Ajal, E.A.; Ouazzani, N.; Fernández-Gutiérrez, A.; Beltrán, G.; Carrasco-Pancorbo, A. First comprehensive characterization of volatile profile of north Moroccan olive oils: A geographic discriminant approach. *Food Res. Int.* **2015**, *76*, 410–417. [CrossRef]

23. Bianchi, F.; Careri, M.; Chiavaro, E.; Musci, M.; Vittadini, E. Gas chromatographic–mass spectrometric characterisation of the Italian Protected Designation of Origin "Altamura" bread volatile profile. *Food Chem.* **2008**, *110*, 787–793. [CrossRef]

24. Baccouri, O.; Bendini, A.; Cerretani, L.; Guerfel, M.; Baccouri, B.; Lercker, G.; Zarrouk, M.; Daoud Ben Miled, D. Comparative study on volatile compounds from Tunisian and Sicilian monovarietal virgin olive oils. *Food Chem.* **2008**, *111*, 322–328. [CrossRef]

25. Sanz, C.; Belaj, A.; Sánchez-Ortiz, A.; Pérez, A.G. Natural Variation of Volatile Compounds in Virgin Olive Oil Analyzed by HS-SPME/GC-MS-FID. *Separations* **2018**, *5*, 24. [CrossRef]

26. Kalua, C.M.; Allen, M.S.; Bedgood, D.R.; Bishop, A.G.; Prenzler, P.D.; Robards, K. Olive oil volatile compounds, flavour development and quality: A critical review. *Food Chem.* **2007**, *100*, 273–286. [CrossRef]

27. Reboredo-Rodríguez, P.; González-Barreiro, C.; Cancho-Grande, B.; Simal-Gándara, J. Concentrations of Aroma Compounds and Odor Activity Values of Odorant Series in Different Olive Cultivars and Their Oils. *J. Agric. Food Chem.* **2013**, *61*, 5252–5259. [CrossRef]

28. Uceda, M.; Frias, L. Harvest dates. Evolution of the fruit oil content, oil composition and oil quality. In Proceedings of the Del Segundo Seminario Oleicola Internacional; IOC: Cordoba, Spain, 1975; pp. 125–128.

29. Brkić Bubola, K.; Koprivnjak, O.; Sladonja, B.; Škevin, D.; Belobrajić, I. Chemical and sensorial changes of Croatian monovarietal olive oils during ripening. *Eur. J. Lipid Sci. Technol.* **2012**, *114*, 1400–1408. [CrossRef]

30. Ranalli, A.; Cabras, P.; Iannucci, E.; Contento, S. Lipochromes, vitamins, aromas and other components of virgin olive oil are affected by processing technology. *Food Chem.* **2001**, *73*, 445–451. [CrossRef]

31. Raffo, A.; Bucci, R.; D'Aloise, A.; Pastore, G. Combined effects of reduced malaxation oxygen levels and storage time on extra-virgin olive oil volatiles investigated by a novel chemometric approach. *Food Chem.* **2015**, *18*, 257–267. [CrossRef] [PubMed]

32. Žanetić, M.; Cerretani, L.; Škevin, D.; Politeo, O.; Vitanović, E.; Jukić Špika, M.; Perica, S.; Ožić, M. Influence of polyphenolic compounds on the oxidative stability of virgin olive oils from selected autochthonous varieties. *J. Food Agric. Environ.* **2013**, *11*, 126–131.

33. Kiritsakis, A.K. Flavor components of olive oil—A review. *J. Am. Oil Chem. Soc.* **1998**, *75*, 673–681. [CrossRef]

34. Brkić Bubola, K.; Koprivnjak, O.; Sladonja, B.; Belobrajić, I. Influence of storage temperature on quality parameters, phenols and volatile compounds of Croatian virgin olive oils. *Grasas y Aceites* **2014**, *65*, e034. [CrossRef]

35. Brkić Bubola, K.; Lukić, M.; Lukić, I.; Koprivnjak, O. Effect of different clarification methods on volatile aroma compound composition of virgin olive oil. *Food Technol. Biotechnol.* **2019**, *57*, 503–512. [CrossRef]

36. Mele, M.A.; Islam, M.Z.; Kang, H.M.; Giuffrè, A.M. Pre- and post-harvest factors and their impact on oil composition and quality of olive fruit. *Emir. J. Food Agric.* **2018**, *30*, 592–603. [CrossRef]

37. Caporaso, N. Virgin Olive Oils: Environmental Conditions, Agronomical Factors and Processing Technology Affecting the Chemistry of Flavor Profile. *J. Food Chem. Nanotechnol.* **2016**, *2*, 21–31. [CrossRef]

38. Vichi, S.; Castellote, A.I.; Pizzale, L.; Conte, L.S.; Buxaderas, S.; López-Tamames, E. Analysis of virgin olive oil volatile compounds by headspace solid-phase microextraction coupled to gas chromatography with mass spectrometric and flame ionization detection. *J. Chromatogr. A* **2003**, *983*, 19–33. [CrossRef]

39. Jukić Špika, M.; Perica, S.; Žanetić, M.; Škevin, D. Virgin Olive Oil Phenols, Fatty Acid Composition and Sensory Profile: Can Cultivar Overpower Environmental and Ripening Effect? *Antioxidants* **2021**, *10*, 689. [CrossRef]

40. Beltrán, G.; Aguilera, M.P.; Del Rio, C.; Sanchez, S.; Martinez, L. Influence of fruit ripening process on the natural antioxidant content of Hojiblanca virgin olive oils. *Food Chem.* **2005**, *89*, 207–215. [CrossRef]

41. Gómez-Rico, A.; Salvador, M.D.; Fregapane, G. Virgin olive oil and olive fruit minor constituents as affected by irrigation management based on SWP and TDF as compared to ETc in medium-density young olive orchards (*Olea europaea* L. cv. Cornicabra and Morisca). *Food Res. Int.* **2009**, *42*, 1067–1076. [CrossRef]

42. Magagna, F.; Valverde-Som, L.; Ruíz-Sambl, C.; Cuadros-Rodríguez, L.; Reichenbach, S.; Bicchi, C.; Cordero, C. Combined untargeted and targeted fingerprinting with comprehensive two-dimensional chromatography for volatiles and ripening indicators in olive oil. *Anal. Chim. Acta* **2016**, *936*, 245–258. [CrossRef]
43. Giuffrè, A.M.; Capocasale, M.; Macrì, R.; Caracciolo, M.M.; Zappia, C.; Poiana, M. Volatile profiles of extra virgin olive oil, olive pomace oil, soybean oil and palm oil in different heating conditions. *LWT* **2019**, *17*, 108631. [CrossRef]
44. Morales, M.T.; Berry, A.J.; McIntyre, P.S.; Aparicio, R. Tentative analysis of virgin olive oil aroma by supercritical fluid extraction-high-resolution gas chromatography-mass spectrometry. *J. Chromatogr. A* **1998**, *819*, 267–275. [CrossRef]
45. Aparicio-Ruiz, R.; García-González, D.L.; Morales, M.T.; Lobo-Prieto, A.; Romero, I. Comparison of two analytical methods validated for the determination of volatile compounds in virgin olive oil: GC-FID vs. GC-MS. *Talanta* **2018**, *187*, 133–141. [CrossRef]
46. Kanavouras, A.; Kiritsakis, A.; Hernandez, R.J. Comparative study on volatile analysis of extra virgin olive oil by dynamic headspace and solid phase micro-extraction. *Food Chem.* **2005**, *90*, 69–79. [CrossRef]
47. Conte, L.; Bendini, A.; Valli, E.; Lucci, P.; Moret, S.; Maquet, A.; Lacoste, F.; Brereton, P.; García-González, D.L.; Moreda, W.; et al. Olive oil quality and authenticity: A review of current EU legislation, standards, relevant methods of analyses, their drawbacks and recommendations for the future. *Trends Food Sci. Technol.* **2019**, *105*, 483–493. [CrossRef]
48. Bianchi, F.; Careri, M.; Mangia, A.; Musci, M. Retention indices in the analysis of food aroma volatile compounds in temperature-programmed gas chromatography: Database creation and evaluation of precision and robustness. *J. Sep. Sci.* **2007**, *30*, 563–572. [CrossRef]
49. Romero, I.; García-González, D.L.; Aparicio-Ruiz, R.; Morales, M.T. Validation of SPME-GCMS method for the analysis of virgin olive oil volatiles responsible for sensory defects. *Talanta* **2015**, *134*, 394–401. [CrossRef] [PubMed]
50. Cecchi, L.; Migliorini, M.; Mulinacci, N. Virgin Olive Oil Volatile Compounds: Composition, Sensory Characteristics, Analytical Approaches, Quality Control, and Authentication. *J. Agric. Food Chem.* **2021**, *69*, 2013–2040. [CrossRef] [PubMed]
51. Procida, G.; Cichelli, A.; Lagazio, C.; Conte, L.S. Relationships between volatile compounds and sensory characteristics in virgin olive oil by analytical and chemometric approaches. *J. Sci. Food Agric.* **2016**, *96*, 311–318. [CrossRef] [PubMed]
52. Cerretani, L.; Salvador, M.D.; Bendini, A.; Fregapane, G. Relationship Between Sensory Evaluation Performed by Italian and Spanish Official Panels and Volatile and Phenolic Profiles of Virgin Olive Oils. *Chem. Percept.* **2008**, *1*, 258–267. [CrossRef]
53. Barbieri, S.; Bendini, A.; Valli, E.; Gallina Toschi, T. Do consumers recognize the positive sensorial attributes of extra virgin olive oils related with their composition? A case study on conventional and organic products. *J. Food Compos. Anal.* **2015**, *44*, 186–195. [CrossRef]
54. Perica, S.; Strikić, F.; Žanetić, M.; Vuletin Selak, G.; Klepo, T. Scientific achievements in Croatian olive sector and future outlook. In Proceedings of the 49th Croatian & 9th International Symposium on Agriculture, Dubrovnik, Croatia, 16–21 February 2014; pp. 7–16.
55. Luna, G.; Morales, M.T.; Aparicio, R. Characterisation of 39 varietal virgin olive oils by their volatile compositions. *Food Chem.* **2006**, *98*, 243–252. [CrossRef]
56. Strikić, F.; Klepo, T.; Rošin, J.; Radunić, M. *Indigenous Cultivars of Olives in Croatia*; Institute for Adriatic Crops and Karst Reclamation: Split, Croatia, 2010; pp. 10–42.
57. International Olive Council. Available online: https://www.internationaloliveoil.org (accessed on 16 September 2021).
58. Ministry of Agriculture. Available online: https://poljoprivreda.gov.hr/maslinarstvo/194 (accessed on 16 September 2021).
59. Lukić, I.; Žanetić, M.; Jukić Špika, M.; Lukić, M.; Koprivnjak, O.; Brkić Bubola, K. Complex interactive effects of ripening degree, malaxation duration and temperature on Oblica cv. virgin olive oil phenols, volatiles and sensory quality. *Food Chem.* **2017**, *232*, 610–620. [CrossRef]
60. Koprivnjak, O.; Brkić Bubola, K.; Majetić, V.; Škevin, D. Influence of free fatty acids, sterols and phospholipids on volatile compounds in olive oil headspace determined by solid phase microextraction-gas chromatography. *Eur. Food Res. Technol.* **2009**, *229*, 539–547. [CrossRef]
61. Žanetić, M.; Škevin, D.; Vitanović, E.; Jukić Špika, M.; Perica, S. Survey of phenolic compounds and sensorial profile of Dalmatian virgin olive oils. *Pomol. Croat.* **2011**, *17*, 19–30.
62. Šarolić, M.; Gugić, M.; Tuberoso, C.I.G.; Jerković, I.; Šuste, M.; Marijanović, Z.; Kuś, P.M. Volatile Profile, Phytochemicals and Antioxidant Activity of Virgin Olive Oils from Croatian Autochthonous Varieties Mašnjača and Krvavica in Comparison with Italian Variety Leccino. *Molecules* **2014**, *19*, 881–895. [CrossRef] [PubMed]
63. Kottek, M.; Grieser, J.; Beck, C.; Rudolf, B.; Rubel, F. World Map of the Köppen-Geiger Climate Classification Updated. *Meteorol. Z.* **2006**, *15*, 259–263. [CrossRef]
64. Gutfinger, T. Polyphenols in olive oils. *J. Am. Oil Chem. Soc.* **1981**, *58*, 966–968. [CrossRef]
65. ISO 12966-2. *Animal and Vegetable Fats and Oils—Gas Chromatography of Fatty Acid Methyl Esters—Part 2: Preparation of Methyl Esters of Fatty Acids*; International Organization for Standardization: Geneva, Switzerland, 2011.
66. ISO 5508. *Animal and Vegetable Fats and Oils—Analysis by Gas Chromatography of Methyl Esters of Fatty Acids*; International Organization for Standardization: Geneva, Switzerland, 1999.
67. Vichi, S.; Guadayol, J.M.; Caixach, J.; López-Tamames, E.; Buxaderas, S. Comparative study of different extraction techniques for the analysis of virgin olive oil aroma. *Food Chem.* **2007**, *105*, 1171–1178. [CrossRef]

68. International Olive Council. *Sensory Analyses of Olive Oil: Method for the Organoleptic Assessment of Virgin Olive Oil*; COI/T.20/Doc. No 15/Rev. 10; International Olive Council: Madrid, Spain, 2018.
69. International Olive Council. *Guide for Selection, Training and Quality Control of Virgin Olive Oil Tasters–Qualifications of Tasters, Panel Leaders and Trainers*; COI/T.20/Doc. No 14/Rev. 7; International Olive Council: Madrid, Spain, 2021.
70. International Olive Council. *Organoleptic Assessment of Extra Virgin Olive Oil Applying to Use a Designation of Origin*; COI/T.20/Doc. No 22; International Olive Council: Madrid, Spain, 2005.
71. Leporini, M.; Loizzo, M.R.; Tenuta, M.C.; Falco, T.; Sicari, V.; Pellicanò, T.M.; Tundis, R. Calabrian extra-virgin olive oil from Frantoio cultivar: Chemical composition and health properties. *Emir. J. Food Agric* **2018**, *30*, 631–637. [CrossRef]
72. Aguilera, M.P.; Beltrán, G.; Ortega, D.; Fernández, A.; Jiménez, A.; Uceda, M. Characterisation of virgin olive oil of Italian olive cultivars: "Frantoio" and "Leccino", grown in Andalusia. *Food Chem.* **2005**, *89*, 387–391. [CrossRef]
73. Velasco, J.; Dobargenes, C. Oxidative stability of virgin olive oil. *Eur. J. Lipid Sci. Technol.* **2002**, *104*, 661–676. [CrossRef]
74. EFSA NDA Panel. Scientific Opinion on the Substantiation of Health Claims Related to Olive Oil and Maintenance of Normal Blood LDL-Cholesterol Concentrations (ID 1316, 1332), Maintenance of Normal (Fasting) Blood Concentrations of Triglycerides (ID 1316, 1332), Maintenan. *EFSA J.* **2011**, *9*, 1–19. [CrossRef]
75. Reboredo-Rodríguez, P.; González-Barreiro, C.; Cancho-Grande, B.; Simal-Gándara, J. Aroma biogenesis and distribution between olive pulps and seeds with identification of aroma trends among cultivars. *Food Chem.* **2013**, *141*, 637–643. [CrossRef] [PubMed]
76. Beltrán, G.; Ruano, M.T.; Jiménez, A.; Uceda, M.; Aguilera, M.P. Evaluation of virgin olive oil bitterness by total phenol content analysis. *Eur. J. Lipid Sci. Technol.* **2007**, *108*, 193–197. [CrossRef]
77. Mayuoni-Kirshinbaum, L.; Porat, R. The flavor of pomegranate fruit: A review. *J. Sci. Food Agric.* **2014**, *94*, 21–27. Available online: https://onlinelibrary.wiley.com/doi/epdf/10.1002/jsfa.6311 (accessed on 24 August 2021). [CrossRef]
78. Lukić, I.; Carlin, S.; Horvat, I.; Vrhovsek, U. Combined targeted and untargeted profiling of volatile aroma compounds with comprehensive two-dimensional gas chromatography for differentiation of virgin olive oils according to variety and geographical origin. *Food Chem.* **2019**, *270*, 403–414. [CrossRef]
79. Caporale, G.; Policastro, S.; Monteleone, E. Bitterness enhancement induced by cut grass odorant (cis-3-hexen-1-ol) in a model olive oil. *Food Qual. Prefer.* **2004**, *15*, 219–227. [CrossRef]
80. Genovese, A.; Yang, N.; Linforth, R.; Sacchi, R.; Fisk, I. The role of phenolic compounds on olive oil aroma release. *Food Res. Int.* **2018**, *112*, 319–327. [CrossRef]

MDPI
St. Alban-Anlage 66
4052 Basel
Switzerland
Tel. +41 61 683 77 34
Fax +41 61 302 89 18
www.mdpi.com

*Plants* Editorial Office
E-mail: plants@mdpi.com
www.mdpi.com/journal/plants